Textbook of

BRYOPHYTES

BASED ON CBCS SYLLABUS OF UNIVERSITY OF DELHI

The Authors

Dr Inderdeep Kaur is an Associate Professor in Botany, Sri Guru Tegh Bahadur Khalsa College, University of Delhi. She has an experience of teaching Archegoniates to the Undergrads for over fifteen years. She has several papers of repute in international and national journals, and has been a participant in conferences in India and abroad. Dr Kaur has authored chapters on various topics in Biology, has co-edited a book entitled "Plant Reproductive Biology and Conservation" and has been Principal Investigator of projects under DST and UGC. Her recent work, a book entitled "Textbook of Gymnosperms" is based on CBCS syllabus and is part of the recommended reading.

Prof. Prem Lal Uniyal is a Professor in Department of Botany, University of Delhi. With specialization in archegoniates, he has been engaged in teaching Post grads. Author of several papers, books and chapters, he has also guided students for Ph.D and is co-author of the book entitled "Textbook of Gymnosperms".

Textbook of
BRYOPHYTES

Based on CBCS Syllabus of University of Delhi

Inderdeep Kaur

Associate Professor
Department of Botany,
Sri Guru Tegh Bahadur Khalsa College,
University of Delhi, Delhi

Prem Lal Uniyal

Professor
Department of Botany,
University of Delhi, Delhi

2020

Daya Publishing House®

A Division of

Astral International Pvt. Ltd.

New Delhi – 110 002

ISBN 9789390371457 (Int. Edition)

Published by : **Daya Publishing House®**
A Division of
Astral International Pvt. Ltd.
– ISO 9001:2015 Certified Company –
4736/23, Ansari Road, Darya Ganj,
New Delhi-110 002
Ph. 011-43549197, 23278134
E-mail: info@astralint.com
Website: www.astralint.com

Preface

Bryophytes comprising three groups – liverworts, mosses and hornworts, are the only land plants with a dominant gametophyte. The persistent gametophyte generation is responsible besides exploratory growth, for proliferation of new generation through sexual, asexual and vegetative processes. Sexual reproduction in this group of plants involves release of motile male gametes into the aquatic environment and requires successful navigation of these naked cells from male to the female sex organs, the archegonia via an external water film. Bryophytes are known therefore, for innovative amphibious nature of survival. The habitats need to be cool, damp and water be available for sexual reproduction. As a result, the members grow luxuriantly in hilly areas where they profusely reproduce. The small size of thallus which is prostrate in most of the genera and exclusive habitat makes them go unnoticed by many of us unless we make a serious effort to reach out to hill slopes or damp soils for a closer look at the vegetation. For this simple reason, students find it difficult to comprehend liverworts, mosses and hornworts as important components of vegetation.

Writing this book has been an endeavour to provide the content of this group of plants with archegonia (other two being – pteridophytes and gymnosperms) in a manner which makes learning simple for students. There are eight chapters in this book. The first chapter focuses on the origin of land plants and role of bryophytes in evolution of land plants. It also gives us the reasons for conserving this flora. Various schemes of classification are discussed and a general plan of the recent classification based on phylogeny (Stotler *et al.,* 2009) is provided in this chapter. The second chapter deals with MARCHANTIOPHYTA – the liverworts, and five genera have been discussed in detail. The third chapter is on BRYOPHYTA – the mosses, with three mosses covered in detail. A detailed account of ANTHOCEROTOPHYTA – the hornworts is given in fourth chapter. The fifth chapter is on role of bryophytes in human welfare which gives us much more than the economic value attached with the group. Sixth chapter is an amalgamation of early theories and recent evolutionary tendencies in bryophytes. From early studies on sex determination in a liverwort, *Sphaerocarpos*, bryophytes have become powerful experimental tools for the elucidation of complex biological processes. The studies in bryology are expanding horizons, with bryophyte molecular genetics unravelling lesser known facts. The recent advances in research on these plants is discussed in the seventh chapter of the book. Studying bryophytes in their natural habitat, and also collecting and processing specimens for research requires some basic knowledge of herbarium techniques which has been dealt with in eighth chapter. To make the laboratory work easy and interesting, colour plates have been added at the end of these chapters. A collection of important questions is given as part of appendix besides an illustrated glossary.

The black and white images complement the text, which is presented in a student-friendly style. After each genus, evolutionary aspects are discussed and a set of questions is provided to check students' knowledge about the group.

Bryophytes have been placed along with Pteridophytes and Gymnosperms under an arbitrary group 'Archegoniates'. Though, a few gymnosperms (*Gnetum* and *Welwitschia*) lack archegonium, but majority of the members have a jacketed archegonium. The systematics of bryophytes has undergone a renaissance following the adoption of techniques, extracting the phylogenetic signal carried by molecular characters. This has broadened the scope of their studies and researchers have tried to workout the relationships between major lineages, besides elucidating the evolution of land plants. Bryophytes are considered to be a pivotal group which holds many clues to early diversification of land plants. A growing consensus suggests that bryophytes represent three separate evolutionary lineages, which have been assigned the rank of division or phylum. With several modern research techniques coming up, data from phylogeny, evolution and molecular characterization, has been integrated to revise the classification. In the present book, we have adhered to higher classification by Crandall-Stotler *et al.* (2009) for PHYLUM MARCHANTIOPHYTA – the liverworts, and by Stotler and Crandall-Stotler (2005) for PHYLUM ANTHOCEROTOPHYTA – the hornworts while PHYLUM BRYOPHYTA – the mosses, has been treated as per the classification of Goffinet *et al.* (2008). Since our previous book – TEXTBOOK OF GYMNOSPERMS dealt with an assemblage of plants which also possess archegonia, it seemed practical to keep consistency of classification in both the books. It is therefore, that the classification scheme by Christenhauz followed for Gymnosperms, has been followed with subclasses under Equisetopsida. It is important for students to note that 'ARCHEGONIATES' is a synthetic group created for the sake of convenience, and which comprises, BRYOPHYTES, PTERIDOPHYTES and GYMNOSPERMS. It is a system which has been followed by University of Delhi, Delhi ever since B.Sc. (Hons) Botany was introduced. However, keeping in view the new trends of classification, our focus is to present representative taxa in light of updated direction of inter-relationships both within and amongst the plants bearing archegonia. The book provides students with revised concepts and emerging trends in bryophyte biology. Special attention has been paid to the format which is not only lucid in style but also easy to understand.

The book is an outcome of thought provoking interactions with students during theory and practical classes. We thank all our students for stimulating our minds to write this textbook.

We are indebted to Dr Surinder Kaur, Associate Professor in Botany, Sri Guru Tegh Bahadur Khalsa College, University of Delhi for her constructive criticism and suggestions which have greatly improved the presentation. We also acknowledge with gratitude the support provided by Prof. Rupam Kapoor, Department of Botany, University of Delhi, Delhi and Dr Monika Koul Moza, Assistant Professor in Botany, Hans Raj College, University of Delhi.

The help received from Ms Garima Anand and Ms Preeti Giri research scholars, Department of Botany, University of Delhi, is duly acknowledged.

We also owe a debt of gratitude to Dr Jaswinder Singh, Principal, Sri Guru Tegh Bahadur Khalsa College, University of Delhi, Delhi and Prof. K. S. Rao, Head, Department of Botany, University of Delhi, for encouraging our scholarly endeavours. We are grateful to Mr. Tanuj for providing us the cover page. Thanks to our publisher ASTRAL International (P) Ltd., New Delhi, for not only their outstanding technical support but also for suggesting intricate modifications that would have never occurred to us.

Inderdeep Kaur

Prem Lal Uniyal

Contents

From Twenty Lessons on British Mosses (1846) by William Gardiner (*Publisher: Longman 1849*)

O! Let us love the silken moss
That clothes the time-worn wall,
For great its Mighty Author is,
Although the plant be small.

The God who made the glorious sun
That shines so clear and bright,
And silver moon, and sparkling stars,
That gem the brow of night.

Did also give the sweet green moss
Its little form so fair;
And, though so tiny in all its parts,
Is not beneath His care.

When wandering in the fragrant wood,
Where pale primroses grow;
To hear the tender ring-dove coo,
And happy small birds sing,
We tread a fresh and downy floor,
By soft green mosses made;
And, when we rest by woodland stream,
Our couch with them is spread.

In valley deep, on mountain high
The mosses still are there;
The dear delightful little things,
We meet them everywhere!

And when we mark them in our walks,
So beautiful, though small,
Our grateful hearts should glow with love
To Him who made them all.

Source

https://digitalcommons.mtu.edu/cgi/viewcontent.cgi?article=1142 and context=bryo-ecol-subchapters

Chapter 1

Introduction

The notion that the biotic world harbours a great range of forms which are easily recognizable and others which are not, extends far back into time when life originated. Despite the range of diversity, the general architecture of plants, essential for survival remains same i.e., roots in advanced forms and rhizoids in algae and/or liverworts; stems in higher plants and fronds in algae, and leaves in higher plants and sporophylls, the spore bearing structures in higher algae/pteridophytes and gymnosperms. In order to assort such diverse plants, attempts have been made to classify them into well-defined groups or taxa based on morphological or phylogenetic characters exhibited by them.

Hornworts, liverworts and mosses commonly referred to as bryophytes, are found throughout the world in variety of habitats – from the harsh environs of Antarctica to the lush green conditions of tropical rain forests. They are considered to be a pivotal group in our understanding of the origin of land plants because bryophytes are believed to be among the earliest diverging lineages of land plants.

According to Chase and Reveal (2009), who have proposed a phylogenetic classification of land plants based on APG III, the Class Equisetopsida has Subclass Anthocerotidae, Subclass Bryidae and Subclass Marchantiidae.

The phylogenetic relationships among the four major lineages of land plants (liverworts, mosses, hornworts and vascular plants) remain vigorously contested.

Commonly referred to along with pteridophytes, as amphibians of plant kingdom (while growing on land, they depend on water for sexual reproduction), bryophytes are believed to have arisen from algal ancestors. While sex organs of bryophytes are provided with a protective jacket, those of algae are not. Bryophytes also share a number of features with pteridophytes and it is for this simple reason that both groups are frequently discussed together.

The points of similarity between bryophytes and pteridophytes are:

- Multicellular sex organs with an outer layer of sterile cells or jacket.
- Permanent retention of the zygote within the female sex organ (archegonium).
- Retention of embryonic sporophyte within archegonium.
- Dependence on water for fertilization.

Both bryophytes and pteridophytes lack complexities of seed bearing plants, the gymnosperms and the angiosperms. Characters specific to highly advanced group of plants, the angiosperms and which are lacking in bryophytes are listed:

- Migration of male gamete through tubular structure (pollen tube).
- Permanent retention of the female gametophyte within the tissues of the sporophyte (ovule/ovary).
- Production of seeds (also related to heterospory).
- Temporary cessation of growth at a certain stage in the development of the embryonic sporophyte (increases survival).

General Account

Bryophytes include liverworts, mosses and hornworts which are usually green, thalloid or leafy plants. They are amongst the simplest of land-dwelling plants (a few are aquatic). The name 'liverwort' is derived from the Anglo-Saxon word 'lifer', meaning liver and 'wyrt', the word for plant. During the 16th century, it was commonly applied to the genus *Marchantia*, a flat, branching, ribbon-shaped plant, the margins of which were claimed to resemble the lobes of a liver. Since the Doctrine of Signatures, a popular concept at the time proclaimed that God had bestowed upon each plant He had created, a mark or sign that pointed to its medicinal value; the liverwort *Marchantia* with its imagined liver-shaped lobes was believed to be useful for the treatment of liver.

Bryophytes is a group of heterogeneous division of the plant kingdom, which includes the three phyla: (i) Marchantiophyta (liverworts, thallose and foliose) with nearly 5000 species in 391 genera (Crandall-Stotler *et al.*, 2009), (ii) Bryophyta (mosses) with about 13000 species (Goffinet *et al.*, 2008), and (iii) Anthocerotophyta (Hornworts) with about 150 species in 14 genera (Stotler and Crandall-Stotler, 2005; Renzaglia *et al.*, 2009). Bryophytes are basically small plants ranging from a few millimeters to a few centimetres. Tallest bryophyte is *Dawsonia* (50-70cms)/*Fontinalis* (3.3 meters), while *Zoopsis* is the smallest bryophyte (5 mm). They possess rhizoids instead of a root system. It may be noted that rhizoids are epidermal cell extensions (while roots in higher plants originate endogenously). The group lacks some of the complex structures seen in other land plants, such as absence of flower or seed formation, no specialized tissue for water conduction and no ways to cut down transpiration because of which the plants cannot tolerate desiccation. They are the amphibians of plant kingdom, spending stages of their life cycle on soil and requiring water for other activities, especially reproduction where flagellate antherozoids have to swim in water film to effect fertilization. Their dependence on water for fertilization imposes restriction on their size and they depend on their immediate surroundings for an adequate supply of moisture. As a result, most bryophytes live in moist, shady or wet locations, growing on rocks, trees, and soil. Growing in clumps and tufts is also a means of water conservation commonly seen in their habit. Some, *e.g., Riccia fluitans, Riella, Fontinalis,* and *Amblystegium* however, have become adapted to aquatic habitat, exhibiting structural differences from the terrestrial forms.

The life cycle shows two phases, the gametophyte and sporophyte. Gametophyte is the most conspicuous, green, long-lived, freely branching and independent phase. The other phase is a short-lived sporophyte, which collapses after the production of spores. The haploid gametophyte is generally thalloid or leafy while the diploid spore-producing sporophyte is small, simple to complex structure. The latter is physiologically dependent upon the independent gamete producing (antheridia and archegonia) gametophyte generation. All bryophytes are homothallic and homosporous with *Marchantia,* one of the few heterothallic species, producing only one type of spore. *Micromitrium* is an extreme example of sexual dimorphism (heterothallism) in which the dwarf male gametophyte grows on the leaves of the markedly larger female plant (Smith, 1955). In *Marchantia polymorpha,* the male and female thalli (vegetative gametophytes) look alike, although reproductive males and females can be distinguished easily by differences in the morphology of the sexual structure (antheridiophore and archegoniophore) each produces. The sex of each haploid gametophyte is determined by cytologically distinct sex chromosomes, with males having one very small Y chromosome and no X chromosome and females having one X chromosome and no Y chromosome (Lorbeer, 1934).

There are two broad categories of liverworts, separated on the basis of their general structure: the thalloid liverworts (Marchantioides) and the leafy liverworts (Jungermannioides). The thalloid forms are flat, dorsiventral, membranous with even, or slightly wavy, lobed margins. The leafy forms exhibit a great diversity in the shape of leaves which may be divided, variously lobed or divided into hair-like segments, or they may be folded into two lobes of unequal size with the smaller lobe situated atop or below the larger (incubous or succubous). Mosses are small plants, at times microscopic which are differentiated into stem and 'leaves' and remain attached to the substratum by rhizoids. Hornworts represent an early land plant group and the common name refers to the elongated horn-like structure – the sporophyte.

Within the three groups, both distinct and specific characters are seen. Moss leaves are of equal size and are spirally arranged. While the main leaves of liverworts are arranged in one plane on either side of the stem, a third row of smaller leaves (amphigastria) is present on the underside of the stem. Leaves of liverworts lack a midrib unlike those of most mosses. Mosses have multicellular, branched rhizoids while liverworts have unicellular unbranched rhizoids. Leaf cells of 90 per cent of liverworts possess two to five oil bodies (as well as numerous chloroplasts) that are absent in moss leaves. The oil droplets in other land plants are just naked bodies of fat but liverwort oil bodies are single membrane-bound organelles that contain essential oils (terpenoids).

Bryophytes are closely related to algae in some features such as thallus-like plant body, lack of vascular tissue, conspicuous gametophyte phase, and retention of the swimming habit by the antherozoids. Early stages of development in gametophyte are represented by green filaments, or protonema which strikingly resemble the filamentous thallus of green algae. The bryophytes and green algae have much in common at the level of cell structure and metabolic pathways. There are however, differences at the tissue level of differentiation. These similarities support the view that bryophytes have evolved from algal ancestors, particularly the green algae.

Bryophytes that share certain characteristics with green algae, also have features in common with Gymnosperms and Pteridophytes, to which a remote common ancestry is suggested (Pickett-Heaps, 1975). For example, the chloroplasts of bryophytes carry a mixture of chlorophylls and carotenoid pigments similar to those found in the other green land plants (and in green algae.) This close resemblance also extends to the ultrastructure of the photosynthetic apparatus, with arrangement of thylakoids and their particulate substructure being virtually identical to that of all other land plants. Bryophytes resemble land plants (and green algae) in important features of their metabolism, such as their food reserves, which consist mainly of starches built of amylase and amylopectin units (Goffinet, 2000; Shaw *et al.,* 2011). The cell walls of bryophytes comprise a structural framework of cellulose microfibrils embedded in a matrix of polygalacturonic acid residues and hemicellulose but, unlike some of the cell walls of vascular plants, lack true lignins (Hebant, 1977).

The spiral form of the motile antherozoids, with their anteriorly located flagella (2 flagella as in bryophytes, pteridophytes like *Selaginella, Lycopodium* and algae like *Chara*, or 25,000, as in Cycads) and posteriorly placed nuclei are features of phylogenetic importance which bryophytes share with a few green algae, pteridophytes and primitive gymnosperms (Pickett-Heaps, 1975; Duckett *et al.,* 1982, 1983). The archegonia, though different in details of structure and development, are similar in their main features to those of the other archegoniates. They provide a clear-cut separation of land plants from the algae. Bryophytes resemble Pteridophytes in some of the features such as the development of embryo within the archegonium protected by the multicellular maternal tissue, presence of stomata, except some liverworts, multicellular sex organs with a jacket of sterile cells and multicellular sporangia/spore producing organs, oogamous sexual reproduction and flagellated male gametes.

Bryophytes play a broadly similar part in the energy and nutrient cycles like those of higher plants. They form the dominant vegetation in many parts of the world where epiphytic mosses and moss peat lands form extensive living biomass. Bryophytes often form an almost continuous carpet on the floor of humid forests. In arctic as well as other extreme environments, bryophytes can produce a substratum that initiates conditions favourable for other plants and fungi to colonize and therefore, contribute to soil building. In such environment, bryophytes are essential in creation of sites in which most other terrestrial organism can survive. The bryophytes are eminently successful plants that can survive under extreme environment. This feature makes wider distribution of bryophytes possible in the world, from artic and alpine environments to the humid tropics; from semi-arid sites to submergence in water. Further, some mosses can recover after many years of nearly complete dehydration. Gametophores of many taxa tolerate periods of freezing, whether under wet or dry conditions.

In certain thalloid liverworts, the gametangia are produced in a raised gametophytic portion called receptacle. The maturation of sex organs is strongly dependent on favourable temperatures and moisture conditions. When antheridia are mature, their rupture is triggered by immersion in water. The archegonia too upon maturity, release fluid residue from the disintegrated neck canal cells when immersed in water. The gametophore in mosses is vital in controlling sporophyte growth and differentiation. The enlarging calyptra influences development and morphogenesis of the sporophyte. The interaction of these two controls is necessary for sporophyte growth and maturation. Calyptra is known to protect apical meristem and physically control movement of materials within the differentiating sporophyte (Uniyal *et al.,* 2007).

Bryophytes are able to absorb water and nutrient ions over most of their surface as they lack cuticle and roots. Water conducting system is highly evolved in some mosses like *Pogonatum* and *Polytrichum* which consists of thick-walled cells (steroids) and in some cases specialized elongated tube-like cells called hydroids. The water economy of most bryophytes resembles that of lichens and terrestrial algae. Bryophytes are poikilohydric plants having a variable water content in living protoplasm which is always near equilibrium with the environment. It means they are metabolically active when water is available and can suspend metabolism to survive in period of drought. In contrast, most vascular plants are homeohydric where water content of the protoplasm varies only within limits even under severe water stress and the plant does not usually have the capacity to survive long periods of drought. The size, structure and water economy of bryophytes makes them to grow in wide range of conditions (Uniyal *et al.*, 2007). Ability of many species to survive periods of severe drought, enables them to colonize apparently inhospitable surfaces of rock, sand and gravel. They also seem well-adapted for life as epiphytes in both, moist temperate and tropical climates. In the cooler parts of the earth, they make an important part of the vegetation and in temperate and cold climates, bryophytes are abundant in fresh water, especially in fast-flowing streams. On the other hand, in lowland tropics bryophytes are scarce or absent in rivers and streams.

Bryophytes are almost lacking in desert and semi-desert regions. Absence of bryophytes in many grassland communities is probably due to competition from higher plants rather than to the physical conditions. Bryophytes do not occur in seas and oceans but a few highly specialized species grow in other salt-rich habitats such as the splash zone above the high tide levels in the sea shores and inland saline soils. *Schistidium maritimum* occurs on maritime rocks (Mukerji & Monoharachary, 2006). The liverwort, *Carrpos* which is found in saltpans in Australia and South Africa is well adapted to saline soils. Many species have the ability to flourish under extremely oligotrophic conditions and are very sensitive to atmospheric and other forms of pollution.

Ecologically, some mosses, hornworts and also liverworts like *Marchantia polymorpha* are considered pioneer plants because they can invade bare areas. Bryophytes particularly mosses, are important as fillers in many ecosystems, closing the plant cover by filling up the spaces between the larger plants. Thus, they check soil erosion, loss of nutrients and also add to the water holding capacity of soil. In the early stages of succession, bryophytes cover bare surfaces, forming a favourable substratum for the establishment of seedlings thus, providing the conditions necessary for the development of subsequent serial stages (Uniyal, 1999). Bryophytes encourage the growth of beneficial microorganisms and provide a congenial habitat for microbial activity and nitrogen-fixing Cyanophyceae (DeLuca *et al.*, 2007). Also, the dead remains of mosses tend to accumulate the nutrients. Contribution of bryophytes to peat formation varies considerably with the ecological conditions. The peat swamps in the tropics, consist mainly of the remains of higher plants, but raised peat of temperate regions may be formed of *Sphagnum* biomass with other acidophyllous bryophytes and flowering plants. *Sphagnum* species are usually the most important peat forming bryophytes, but in parts of Southern Hemisphere and tropical mountains, *Breutelia* species contributes to peat formation. Peat formation is only one of the many ways in which bryophytes modify their surroundings (Hayward, 1982). Bryophytes play an important role in modifying their environment and affecting the life of other organisms.

Land Habit

Bryophytes are known as 'amphibians' of the kingdom Plantae. Most of the bryophytes are land dwellers inhabiting damp, shaded and humid sites. A few of them, however, occur in or float in water and the aquatic habit has been acquired secondarily by these plants. When water dries up, bryophytes still grow equally well on the drying mud. They can withstand long periods of drought. During this period, they become almost brittle but with the onset of rainy season, the apparently dried, and brittle thalli, turn green ready to carry out normal life functions.

Bryophytes have evolved from aquatic ancestors like algae. In order to adapt to land habit, they have developed several features:

- The thick compact, multicellular, thallus has reduced total surface area of a compact body in proportion to its volume which protects it against desiccation.

- They have rhizoids with important functions of absorbing water from soil through endo- and exo-conduction and anchoring the plants in soil.
- Specialized tissues – hydrome and leptome are developed for conduction of water minerals and photosynthates in more advanced forms.
- Sex organs are multicellular and jacketed. The fertilized egg is retained in the archegonium and develops into embryo inside the jacketed archegonium and later acquires additional layers for protection such as calyptra.
- For efficient gaseous exchange, liverworts frequently develop air pores on the upper surface of the thallus while mosses develop functional stomata in the sporophytic tissues.
- Spores are thick-walled and wind disseminated.
- Habit to grow in clumps and tufts, protects them from dessication and also provides mechanical support to individual thallus to grow above the ground.
- Efficient means of vegetative reproduction also helps in spread of the species.
- The evolution of bryophytes occurred through development of the heterotrichous habit (as seen in *Coleochaete*) followed by development of parenchymatous tissue in upright portion and disappearance of prostrate system.
- Finally, during further evolution, plants also acquired a primitive conducting strand and a cuticularized surface which helped in land habit.

**Evidences – like chloroplast genome sequence matrix, maximum parsimony and compatibility methods, strongly support liverworts as the sister to all other land plants and hornworts as the sister to vascular plants.*

Two major events occurred in plant evolution, 1) water-to-land transition, 2) change from a haploid gametophyte dominant life cycle in bryophytes to diploid sporophytic generation, dominant in life cycle of vascular plants.

Salient Features

- The bryophytes are a group of primitive land dwellers with a thalloid or leafy, small green plant body.
- In their vegetative structures, they have become entirely adapted to the land habitat. However, for the sexual reproduction (transfer of antherozoids from antheridia to archegonia, dehiscence of sex organs and capsules) bryophytes depend on water. This dependence is also seen in vegetative reproduction, if carried out through structures such as gemmae.
- Bryophytes lack the lignified vascular tissues and in highly evolved mosses internal conduction is of the primitive type.
- The sexual reproduction is of oogamous type and the sex organs are jacketed archegonia as female sex organs and antheridia as male.

- The antherozoids are biflagellate, whiplash type.
- The plant body lacks true roots, stem or leaves. In lower bryophytes (liverworts) plant body is flat which grows prostrate and attached to substratum by means of delicate, unbranched, unicellular hair-like rhizoids. While in higher bryophytes (mosses), the plant body is erect which includes central axis bearing leaf-like expansions. It is attached to substratum by branched, multicellular rhizoids.
- In the life cycle, alternation of generation is observed with a conspicuous and dominant phase, gametophyte (main plant body), which is independent and related with sexual reproduction. Sporophyte, the spore bearing phase is physiologically dependent on the gametophyte for its nutrition.
- The sporophyte (capsule) is associated with the production of wind dessiminated non-motile, cutinized spores which are a product of meiosis.
- The dehiscence of capsule occurs in a highly specialized manner and involves various tissues in advanced forms such as *Funaria.*

Classification

Linnaeus (1753) was the pioneer to recognize genus Jungermannia to comprise both leafy and thalloid taxa and he placed 25 species under this genus. These 25 species were then revised to 21 species by Gray (1821), and Corda (1829) respectively. Gradually as more species were discovered, new systems of classification were proposed. Notable amongst these is classification of Endlicher (1841) who adopted order name Hepaticae (which was originally proposed by de Jussieu, (1789). Later, Schiffner (1893) divided Jungermanniales into subgroups - akrogynae and anakrogynae. Major work of Endlicher (1841), Campbell (1891) and Cavers (1910-1911) led to ascending series of taxa based on the fact that sporophytes evolved from simple to complex structures by progressive elaboration of sterile tissues (Bower, 1890). The other school of thought however, believed in reductive evolution (Church, 1919).

In one of the other schemes of classification, Braun (1864) was the first to coin the term Bryophyta or acotyledonae for a group of Algae, Fungi, Lichens and Mosses. Schimper (1879) treated Bryophyta as a distinct division of the plant kingdom while Eichler (1883) divided Bryophyta into two groups Hepaticae and Musci. Engler (1892) divided Hepaticae into three orders: Marchantiales, Jungermanniales and Anthocerotales, and Musci also in three orders: Sphagnales, Andreaeales and Bryales. This classification was followed by a number of workers. Subsequent schemes of classification became more practical and a significant attempt was made by Bold (1956) who gave the status of divisions to the class Hepaticae and Musci- thus, they were called Hepatophyta and Bryophyta respectively.

Anthocerotales usually listed as an order under Hepaticae was given an isolated rank of a class by Underwood (1894) and Gayet (1897). Later, Takhtajan (1953) distinguished three classes in Bryophyta: Hepaticae, Anthocerotae and Musci.

Rothmaler (1951) changed the nomenclature of three classes to Hepaticopsida, Anthocerotopsida, and Bryopsida.

Within class – Hepaticopsida, Cavers (1910), Evans (1939) and many others considered three orders— Marchantiales, Jungermanniales and Sphaerocarpales. Campbell (1935) added Calobryales and later when *Takakia* was discovered in 1951, Cavers (1910), Campbell (1935) and Schuster (1963) recognized 5 orders in Hepaticopsida:

1. Takakiales
2. Calobryales
3. Jungermanniales
4. Marchantiales
5. Sphaerocarpales

Class Anthocerotopsida had a monotypic order Anthocerotales with a single family Anthocerotaceae while Bryopsida had 5 subclasses:

1. Sphagnidae
2. Andreaeidae
3. Polytrichidae
4. Buxbaumiidae
5. Bryidae

The three classes: Hepaticae (Liverworts and Scale mosses), Anthocerotae (hornworts) and Musci (mosses) were renamed based on International Rules of Nomenclature which recommended them as Hepaticopsida, Anthocerotopsida and Bryopsida (Schofield, 1985).

Classification of bryophytes which remained for greater period primarily based on morphological evidences, such as gametophytic and sporophytic structures, position of gametangia, leaf cell patterns and peristome characters gradually became supplemented with the molecular data. The inter-relationships between various taxa, phylogeny and evolution biology have led to new understanding of taxa, and have called for new classification schemes. The classification schemes proposed by Crandall-Stotler and Stotler (2000), Goffinet and Buck (2004) and Duff *et al* (2007) reflect the state of diversity of such a simple plant group. According to the authors, about two hundred years of bryological investigations of morphological, anatomical and developmental characters formed the foundation for systematics which was followed for several years by researchers all over the world. With the development of phylogenetic theory and techniques allowing DNA extraction and amplification and sequencing of specific loci, new information has become available, which has been used to revise classification of Bryophytes.

*The classification presented is based on thorough revisions of Crandall-Stotler *et al.* (2009) for Marchantiophyta (liverworts), Goffinet *et al.* (2008) for

Bryophyta (mosses) and Stotler and Crandall- Stotler (2005) for Anthocerotophyta (hornworts). The inter-reationships of the groups are discussed by Renzaglia *et al.* (2009). Considering the phylogenetic inter-relationships between various groups of plants, it has become evident that bryophytes are the first green plants to successfully occupy various terrestrial niches (Shaw & Renzaglia, 2004).

The recent classification is entirely a new concept based on phylogeny and cladistics, and is quite elaborate. However, for simplicity of undergraduate understanding, it has been summarized here, with a brief classification of genera included in syllabi.

Since only selected genera are included as type study, detailed classification of liverworts and mosses is not provided here. However, classification of *Marchantia, Riccia, Pellia, Porella* and *Frullania* (Liverworts), *Sphagnum, Polytrichum* and *Funaria* (Mosses) which are included in syllabus is provided in the following pages. For hornworts however, a detailed classification is provided as the group is monotypic.

Liverwort – Classification at the Rank of Genus and Above

PHYLUM **MARCHANTIOPHYTA** Stotler and Crand.-Stotl., in A. J. Shaw and B. Goffinet, Bryoph. Biol.: 63 (2000).

CLASS **MARCHANTIOPSIDA** Cronquist, Takht. and W. Zimm., Taxon 15: 132-133 (1966). ("Marchantiatae").

SUBCLASS **MARCHANTIIDAE** Engl. [Unterklasse "Marchantiales"] in: A. Engler and K. Prantl, Nat. Pflanzenfam. I (3): 1 (1893).

ORDER **MARCHANTIALES** Limpr. in Cohn, Krypt.-Fl. Schlesien 1: 239, 336 (1877).

FAMILY **MARCHANTIACEAE** Lindl., Nat. Syst. Bot. (ed. 2): 412 (1836). *Bucegia* Radian, ***Marchantia*** L., *Preissia* Corda

FAMILY **RICCIACEAE** Rchb., Bot. Damen: 255 (1828).

Riccia L., *Ricciocarpos* Corda

CLASS **JUNGERMANNIOPSIDA** Stotler and Crand.-Stotl., Bryologist 80: 425 (1977).

SUBCLASS **PELLIIDAE** He-Nygrén, Juslén, Ahonen, Glenny and Piippo, Cladistics 22: 27 (2006).

ORDER **PELLIALES** He-Nygrén, Juslén, Ahonen, Glenny and Piippo, Cladistics 22: 27 (2006).

FAMILY **PELLIACEAE** H. Klinggr, Höh. Crypt. Preuss.: 13 (1858).

This classification is taken from: Crandall-Stotler, B., R. E. Stotler and D. G.Long. [2008]2009. **Morphology and classification of the Marchantiophyta, pp. 1-54. In B. Goffinet and A. J. Shaw (eds.) Bryophyte Biology, 2nd edition. Cambridge University Press, Cambridge. Modifications have been made based upon the more recent publication: Crandall-Stotler, B., R. E. Stotler and D. G. Long. 2009. **Phylogeny and Classification of the Marchantiophyta**, Edinburgh Journal of Botany 66: 155-198.*

Noteroclada Taylor ex Hook. and Wilson, ***Pellia*** Raddi nom. cons.

SUBCLASS **JUNGERMANNIIDAE** Engl. [Unterklasse "Jungermanniales"] in A. Engler and K. Prantl, Nat. Pflanzenfam. I (3): 1 (1893).

ORDER **PORELLALES** Schljakov, Bot. Zhurn. (Moscow and Leningrad) 57: 505 (1972).

SUBORDER **PORELLINEAE** R. M. Schust., J. Hattori Bot. Lab. 26: 229 (1963).

FAMILY **PORELLACEAE** Cavers nom. cons., New Phytol. 9: 292 (1910).

Ascidiota C. Massal., ***Porella*** L. [including *Macvicaria* W. E. Nicholson]

SUBORDER **JUBULINEAE** Müll. Frib., Lebermoose 1: 403 (1909).

FAMILY **FRULLANIACEAE** Lorch in G. Lindau, Krypt.-Fl. Anf. 6: 174 (1914).

Frullania Raddi [including *Amphijubula* R. M. Schust., *Neohattoria* Kamim., *Schusterella* S. Hatt., Sharp and Mizut., and *Steerea* S. Hatt. and Kamim.]

Moss – Classification at the Rank of Genus and Above

(modified from Goffinet, B., W.R Buck and A.J. Shaw, Bryophyte Biology, 2nd ed. Cambridge University Press, 2008; Goffinet 2012: http://www.eeb.uconn.edu/people/goffinet/Classificationmosses.html)

PHYLUM **BRYOPHYTA** (Selected genera only)

SUBPHYLUM **SPHAGNOPHYTINA** Doweld

CLASS **SPHAGNOPSIDA** Ochyra

ORDER **SPHAGNALES** Limpr

FAMILY **SPHAGNACEAE** Dumort.

Sphagnum L.

SUBPHYLUM **BRYOPHYTINA** Engler

CLASS **POLYTRICHOPSIDA** Doweld

ORDER **POLYTRICHALES** M Fleisch

FAMILY **POLYTRICHACEAE** Schwagr.

Pogonatum P. Beauv., ***Polytrichum*** Hedw.

CLASS **BRYOPSIDA** Rothm

SUBCLASS **FUNARIIDAE** Ochyra

ORDER **FUNARIALES** M. Fleisch.

FAMILY **FUNARIACEAE** Schwagr.

Physcomitrium (Brid.) Brid., ***Funaria*** Hedw.

Hornwort – Classification at the Rank of Genus and Above

Reference: Stotler, R. E. and B. Crandall-Stotler. 2005. A revised classification of the Anthocerotophyta and a checklist of the hornworts of North America, north of Mexico. Bryologist 108: 16-26.

PHYLUM **ANTHOCEROTOPHYTA** Stotl. and Crand.-Stotl. Bryologist 80: 425. 1977. (A DETAILED ACCOUNT)

CLASS **LEIOSPOROCEROTOPSIDA** Stotl. and Crand.-Stotl., Bryologist 108: 24. 2005

ORDER **LEIOSPOROCEROTALES** Hässel, J. Hatt. Bot. Lab. 64: 82. 1988.

FAMILY **LEIOSPOROCEROTACEAE** Hässel, J. Bryol. 14: 255. 1986.

Leiosporoceros Hässel, J. Bryol. 14: 255. 1986.

CLASS **ANTHOCEROTOPSIDA** Jancz. ex Stotl. and Crand.-Stotl., Bryologist 108: 24. 2005.

ORDER **ANTHOCEROTALES** Limpricht in Cohn, Krypt. Fl. von Schlesien: 239, 345. Breslau. 1877.

FAMILY **ANTHOCEROTACEAE** Dumort., Analys. Fam. Pl. 68-69. Tournay. 1829.

Anthoceros L., Sp. Plt.: 1139. Stockholm. 1753.

Folioceros D. C. Bharadwaj, Geophytology 1: 9. 1971.

Sphaerosporoceros Hässel, J. Hattori Bot. Lab. 64: 79. 1988

ORDER **NOTOTHYLADALES** Hyvönen and Piippo, J. Hatt. Bot. Lab. 74: 117. 1993. "Notothylales".

FAMILY **NOTOTHYLADACEAE** (Milde) Müll. Frib. ex Prosk., Phytomorphology 10: 10. 1960.

SUBFAMILY **NOTOTHYLADOIDEAE** Grolle, J. Bryol. 7: 215. 1972.

Notothylas Sull. ex A. Gray, Amer. J. Sci. Arts, ser. 2, 1: 74. 1846.

SUBFAMILY **PHAEOCEROTOIDEAE** Hässel, J. Hatt. Bot. Lab. 64: 81. 1988.

Hattorioceros (Hasegawa) Hasegawa, J. Hattori Bot. Lab. 76: 32. 1994.

Mesoceros Piippo, Acta Bot. Fenn. 148: 30. 1993.

Phaeoceros Prosk., Bull. Torrey Bot. Cl. 78: 346. 1951.

Phymatoceros Stotler, W. T. Doyle and Crand.-Stotl., Phytologia 87: 113. 2005.

FAMILY **DENDROCEROTACEAE** (Milde) Hässel, J. Hatt. Bot. Lab. 64: 82. 1988.

SUBFAMILY **DENDROCEROTOIDEAE** R. M. Schust., Phytologia 63: 195, 200.

Dendroceros Nees in Gottsche, Lindenberg and Nees, Syn. Hep.: 579. 1846.

SUBFAMILY **MEGACEROTOIDEAE** Stotl. and Crand.-Stotl., Bryologist 108: 24. 2005.

Megaceros Campb., Ann. Bot. (London) 21: 484. 1907.

Nothoceros (R. M. Schust.) Hasegawa, J. Hatt. Bot. Lab. 76: 32. 1994. "*Nothoceros*"

Vegetative Reproduction

The vegetative reproduction in bryophytes is a very efficient method, profoundly affected by climatic conditions (Schofield, 1985).

Fragmentation: The older portion of the thallus or shoots near the branching decay and die by natural or mechanical means, leading to the separation of younger branches. The separated segments then form individuals.

Adventitious branches: These branches arise from the underside of the thallus or base of the shoots, which on separation, form new plants. The adventitious branches readily detach to serve as effective diaspores in adverse conditions.

Innovations: The lateral branches which appear in the vegetative plant in form of comal tuft, disjoin from the parent plant and grow into new individuals as in *Sphagnum*.

Cladia: Cladia are small branches which arise either from the leaf (*Bryopteris, Plagiochila, Bazzania, Frullania*) or from the stem (*Leptolejeunea, Drepanolejeunea*) and give rise to new individuals.

Tubers: At the end of growing season, special subterranean branches arise, which store food and become swollen. These branches (tubers) develop into new individuals in favourable conditions. The tubers are common in most of the thalloid forms which cope with desiccation.

Separation of vegetative buds: The vegetative buds of the modified branches are shed and serve the purpose of regeneration.

Gemmae: Globose discoid plate-like structures are produced on the thallus, leaf tips or modified stem apices. Multicellular stalked discoid gemmae are produced in gemma cups on the dorsal surface of the thallus of *Marchantia, Lunularia* and *Cavicularia*. Two-celled endogenous gemmae are formed within the superficial cell of *Riccardia* and *Haplozia*. On the leaves of *Radula, Cololejeunea, Colura* and *Leptocolea,* discoid multicellular gemmae are produced. Three or four-celled gemmae are produced in the axil of the leaves of *Treubia*, while star shaped gemmae are produced in the thallus of *Blassia*. In *Anthoceros,* gemmae are borne on the dorsal surface of the thallus along the margins. Gemmae are borne on highly specialized receptacles in *Tetraphis, Aulacomnium* and *Calymperes*. Globular multicellular gemmae arise at the base of the leaf in *Bryum*. In many taxa of Pottiaceae, (early classification), the gemmae arise on the rhizoids. Gemmae germinate in favourable conditions and develop new gametophores.

Secondary Protonema: Filamentous structure formed from the living cell of gametophore other than spores is called secondary protonema. In *Funaria* and *Sphagnum,* secondary protonema develop from rhizoids and give rise to erect leafy gametophores.

Alternate Pathways of Life Cycle

Bryophytes have remarkable regeneration potential. Any fragment of the plant (gametophytic or sporophytic) regenerates easily in favourable conditions and is able to form a complete plant. The sporophyte arises directly from the ordinary cells of the gametophyte, without the participation of gametes and zygote formation. The apogamous sporophytes grow exposed from the very start, there being no archegonial covering or calyptra (Chopra & Kumra, 1988).

The occurrence of apogamous sporophytes in bryophytes was first reported by Springer (1935) on the leaf and stem tips of naturally growing diploid gametophytes of the moss *Phascum cuspidatum*. Apogamous sporophytes could be induced regularly and predictably at sites where gametophytes normally developed. The diploid gametophytes of *Pottia intermedia* were obtained by regeneration from the sporophyte. Numerous apogamous sporophytes with well- developed capsules appeared on the gametophytes. Some of these sporophytes produced viable haploid spores by regular meiosis (Ripetsky & Metasov, 1973).

Physcomitrium pyriforme remains an elegant system for studying apogamy at different levels (Menon & Lal, 1972, 1974, 1977). The haploid gametophytes and secondary protonema when cultivated in sugar free-medium produce gametophytic buds. When these buds are transferred to a medium supplemented with 2 per cent sucrose, the protonema produces a large number of apogamous sporophytes.

In bryophytes, the development of gametophytes directly from ordinary vegetative cells of the sporophyte is also a common phenomenon. The steps of meiosis and spore formation are bypassed. The entire seta of moss sporophyte or a segment, is capable of producing a filamentous protonema *in vitro*. The protonema produced by regeneration from the seta ramifies in the normal manner and eventually produces green leafy gametophytes. The ability to regenerate aposporously has been reported in many liverworts. The first record of apospory was in the Jungermanniales where gametophytes were obtained from over 90 per cent of the seta from sporophytes at the spore mother cell stage (Raudzens & Matzke, 1968). Apospory has been recorded in other bryophytes like *Pellia epiphylla, Pallavicinnia lyelli* (Hook), *Marchantia* sp. (Burgeff, 1943) and in *Anthoceros levis* (Schwarzenbach, 1926).

Ecology

Bryophytes have high water retention capacity due to their aggregated growth form and tendency to be most abundant in regions with high levels of atmospheric humidity and low rates of evaporation. They can quickly absorb water, release it slowly into the surrounding environment, contributing to the retention of humidity in a forest and regulation of water flow. Bryophytes are important in stabilizing the soil crust of habitats such as steep and sloping banks in woodland. They trap and accumulate soil and nutrients in the patches of gametophores (Leach, 1931). In some semi-arid woodlands, bryophytes play important roles as colonizers and soil stabilizers in areas where soil surface conditions have declined as a result of increased infiltration (Schofield, 1985). Extended colonies of mosses prevent soil erosion in open sites and disturbed habitats. Bryophytes also provide suitable substrates for the growth of blue-green algae, bacteria, fungi and small animals (Rabatin, 1980; Govindayari *et al.,* 2009).

Bryophytes form more of the photosynthetically active biomass in moist montane forests than all the other plant groups together and therefore, play important roles in the hydrological, chemical, and organic matter cycling. The ectohydric species receive much of their nutrient supply from water running over the substrate whereas forest-floor bryophytes often acquire most of their nutrients from canopy through fall. Many minerals are available through both leachates and dust accumulated on the trees and washed down, during rainstorms to the forest floor. From the leaves of the trees, P and K are readily leached, while Ca and Fe are often abundant in dust, and the faeces of insect larvae contribute N and P. Bryophytes nearest to the tree base have appeared to be the most favorably located for nutrient availability and uptake, while more distant from the tree base, there is a marked decline in element uptake (Tooren, 1988; Bahuguna *et al.,* 2012).

Responses to Pollution

Pollutants inhibit sexual reproduction in bryophytes. They reduce photosynthesis and growth of plants and may eventually cause their death. Some species of bryophytes have become rare as a result of increasing pollution. Increasing precipitation enhances the heavy metal content in bryophytes (Tyler, 1971, Ghatge *et al.*, 2011). The rough mats, tall turfs, large cushions and leafy liverworts are least resistant to pollutants. The protonema of mosses are more sensitive than the mature gametophores. Sensitive taxa are used as indicators, however tolerant taxa can be used accumulators.

SO_2 and NO_2 are two gases that get converted into strong acids to form acid rain which causes maximum damage to bryophytes. An increased acidity may damage cell membranes, solubilize potentially toxic metals like Al^{3+} and worsen the impact

Do You Know?

Various invertebrates (mites, springtails, midges, leaf-hoppers and aphids) visit the antheridia-bearing rosettes of *Polytrichum commune.* The invertebrates feed on the mucilage produced by the paraphyses found in antheridial heads. While rummaging amongst the antheridia, the invertebrates get smeared with mucilage in which are mixed viable antherozoids. In experiments with the moss *Bryum capillare,* female plants two metres away from the nearest male could be fertilized when insects visited the female mosses. However, there was no fertilization when the female plants were covered with a very fine net that excluded insects (https://www.anbg.gov.au/bryophyte/sex-sperm-dispersal.html).

of other pollutants like SO_2 (Farmer *et al.,* 1992). High deposition of nitrogen in some areas has led to a decline of the montane *Racomitrium* heaths. Ammonia is highly phytotoxic and causes leaf tip chlorosis followed by necrosis as reported in *Racromitrium lanuginosum, Dicranum spurium* and *Campylopus flexuosus*. Bryophytes showed impairment of photosynthesis and/or increased membrane leakage when subjected to acute ozone exposures (Lee *et al.,* 1998). Mercury is particularly toxic, low concentration greatly inhibits photosynthesis, temporarily increases respiration, reduces chlorophyll levels and causes loss of intracellular K^+ as from *R. squarrosus* (Brown & Whiteland, 1986). Aquatic bryophytes are affected by acidification, increase in nutrients, heavy metals and radionuclide as well as by organic pollutants, oxygen deficiency, turbidity and temperature increases.

Bryophyte Communities

Epiphytes

Many bryophytes grow on the bark of the trees and shrubs. They either form cushions or are adpressed to bark or hang down from the branches of trees. They also thrive over exposed roots of trees. Epiphytic habitats are free of competition while rough and moist bark promotes the growth of bryophytes. Epiphytic taxa take water from the stem through fall. Some obligate epiphytic taxa that grow exclusively on tree bark are *Tortula laeviphila, Tortula papillose, Zygodon conoideus, Orthotrichum, Ulota, Meteorium, Meteoriopsis, Cryptoleptodon* and *Aerobryopsis*. The taxa on the top branches of the tree are highly tolerant to desiccation, however the taxa on the tree base are desiccation sensitive (Negi & Gadgil, 1997). Moss growth on the tree bark is beneficial as mosses keep congenial environment.

Epiphylls

These bryophytes are short-lived and to be reckoned among the biological nomads of the forests. They grow on the surface (usually the upper) of leaves of evergreen trees especially in the tropical and humid temperate region (Negi & Gadgil, 2001). Mosses and foliose hepatics such as *Crossomitrium, Chaetomitrium sestosum, Demorphocladon borneense* and *Taxithelium* are the conspicuous epiphyllous plants. They spread over the surface of the leaf but do not take any resources from the leaf.

Epiliths or Saxicolous

Many bryophytes spread over the smooth surface of rocks. The gametophores grow in compact turf form or mats. They are highly desiccation tolerant. Some of the obligate epiliths are *Tortula, Gymnostomum, Grimmia, Seligeria, Blindia, Porella, Marchantia mackaii, Frullania microphylla.* Facultative epiliths are *Dicranum, Tortula tortuosa, Bryum capillare, Heteropterum, Isothecium, Brachythecium, Homolothecium sericeum, Hypnum cupressiforme, Diplophyllum albicans, Porella platyphylla* and *Frullania*.

Desert Bryophytes

Bryophytes grow in the shaded and moist sites in desert. The gametophores of desert bryophytes grow in tufts and clumps to reduce water loss and trap moisture during night. Thalloid taxa of dry habitats have air chambers for efficient gas exchange. A few thalloid taxa of *Plagiochasma* sp, *Athalamia, Mania, Oxymitra* and *Exormotheca pustulosa* occur in drier sites. Mosses are more tolerant to dry conditions and *Barbula torquata, Tortula princes, T. ruralis, Bryum pendulum* and *Ceratodon purpurens* are found in deserts. Bryophytes generally prefer to grow in those desert soils which are moderately rich in minerals especially salt, lime ($CaCO_3$) and gypsum.

Taxa on Disturbed Soil

A characteristic community of bryophytes is found on fresh soil, termites, and nests as well as on road cuttings, vehicles tracks and other artificially disturbed ground (Grytnes *et al.,* 2006). These communities consist of short-lived tolerant pioneer taxa. The spores of these taxa germinate on the disturbed land and form extensive protonema and buds. The disturbed site bryophytes are *Fissidens taxifolius, Funuria, Timmiella, Barbula, Tortula, Hyophila* and *Ceratodon*. They predominantly include cleistocarpous or protonemal mosses. *Physcomitrium, Trematodon, Bryum* and *Oxystegus* occupy disturbed sites. Large thalloid mats of *Marchantia, Plagiochasma* and *Asterella* and cushions of *Philonotis* and *Bryum, Leucobryum* sps appear on roadside banks and road cuts.

Epixylic

These bryophytes occur mainly on dead woods. The fallen logs are initially softened by fungi and then the bryophyte spores easily germinate on the moist wood. Pleurocarpous mosses such as *Sematophyllum subsimplex, Taxithelium planum,* and some acrocarpous species such as *Leucobryum maritime, Octoblepharum albidum* and *Calymperus lonchophyllum* mainly occur as a carpet form on dead woods.

Calcicoles and Calcifuges

Calcicoles are calcium loving and calcifuges include calcium averting taxa. Calcicoles are restricted to rocks and soils containing calcium carbonate or inhabit waters that flow over or percolate through calcium-rich substrata. Calcium precipitates over the mosses and there is a formation of calcium rock called tufa. Some of the calcifuges are *Andreaea rothii, Marsupella emarginata* and *Scapania undulate*. Calcicole communities are mainly composed of *Grimmia anodon, Pseudoleskeela catenulate, Schistidium apocarpum, S. trichodon, S. atrofuscum, Philonotis, Fissidens, Hydrogonium, Gymnostomum, Bryoerythrophyllum, Encalyptra, Bryum argenteum* and *Tortula mucronifolia.*

Post Fire Community

Burnt areas provide habitats for a succession of mosses. The fires deposit large amount of ashes raising the surface temperature upto 90°C that greatly modify soil and the pH rises with the accumulation of high levels of inorganic elements, various forms of nitrogen, PO_4 and soluble organic matter. Such sites favour the growth of taxa that require high amount of inorganic elements and high pH. The spores which land on such sites, easily form pioneer colonies. *Funaria hygrometrica, Ceratodon purpureus, Bryum argenteum* and other *Bryum* species form the pioneer moss community in burnt sites.

Cave Bryophytes

Bryophytes may adjust themselves right from the orifice upto the interior of the dolomite caves. These taxa are adapted to survive in low light conditions. *Targionia hypophylla, Plagiochila chinensis, Porella densifolia, Anomodon rugelli, Brachythecium neckeroideum* and *Thuidium tamariscellum, Ephemerum, Cyathodium, Cryptomitrium himalayense, Lejeunea, Fissidens, Isopterygium albescens, Plagiothecium neckeroideum,* occur in the cave mouth where light remains available for a short period.

Coprophilous Bryophytes

Some bryophytes normally inhabit decomposed animal excrement or carcases. Some of the coprophilous mosses include *Tayloria rudolphiana, Tetraplodon* and members of Splachnaceae which grow strictly on cow dung or other animal dung. Their spores are dispersed by flies.

Halophytes

Very few species of bryophytes are true halophytes and associated with salt spray and coastal dune. *Schistidium maritimum, Ulota phyllantha* and *Tortula flavorensis* found to grow in the splash zones of sea shores, appear to be highly salt tolerant.

Physiology

Bryophytes obtain most of their water and minerals from their surroundings through diffusion along the entire surface. They show great variation in their ability of water uptake and movement. In endohydric species, water is taken up from the underlying substratum and conducted internally through specialized conducting cells called hydroids, to the leaves (Polytrichaceae, Mniaceae, *Funaria hygrometrica, Bryum* sp) (Buch, 1945). Some endohydric bryophytes have underground rhizomes (*Dawsonia, Polytrichum, Climacium*) or root-like structure (*Haplomitrium, Takakia*), which help in absorbing and retaining water. Rhizoids provide intimate contact between the stem base and the soil surface as well as form cable-like structures as in mosses with many capillary spaces that help in making continuous stream of water and nutrients as in mosses. In ectohydric bryophytes, water is readily absorbed over the entire surface and water movement is much more diffuse as seen in Grimmiaceae, Orthotrichaceae and all pleurocarpous mosses and leafy liverworts.

Species that combine external conduction with predominantly internal conduction are called mixohydric. Members of Bryidae, Marchantiales and several Metzgeriales are mixohydric. Steriodal cells of the stem, which form a prominent cortex in the taxa of dry habitats, also conduct water. Water uptake is relatively rapid as also the water loss. However, the morphology allows cut down of rapid water loss of such as dry, colorless hair points on leaves of some mosses (*Grimmia*) inhibit water loss by forming a boundary layer on the surface of the gametophores where evaporation does not take place. Tight turfs and cushions in habit retain moisture more effectively than loose turfs or wefts. These growth forms characterize the water regimes of sites and are useful as indicators of microenvironment. Many mosses and leafy hepatics show an extraordinary high tolerance to extreme desiccation and resume normal metabolism very rapidly after rehydration. *Tortula ruralis* occurs in the drier site and tolerates drought for longer periods (10 months) and recovers to normal photosynthesis within a few hours of rehydration (Bewley, 1974).

*Distribution of Bryophytes in India (Survey Based on Early Classifications)

Nearly 2500 species of bryophytes are found in India. The Indian bryo flora is a confluence of the flora of the adjacent areas. They are mostly confined to Western and Eastern Himalayas and Western Ghats (Chopra, 1975; Gangulee, 1976; Daniels & Kariyappa, 2007; Dandotiya *et al.*, 2011; Govindapyari *et al.*, 2012,). These areas provide suitable conditions for the luxuriant growth of bryophytes.

The checklist of the bryophytes of India reports total 2489 taxa of bryophytes from India, comprising 1786 species in 355 genera of mosses, 675 species in 121 genera of liveworts and 25 species in 6 genera of hornworts. Some of the genera of mosses like *Fissidens, Barbula, Campylopus* and *Bryum* are found to have largest number of species. In liverworts, *Riccia, Porella, Frullania, Lejeunea, Plagiochila* and *Jungermannia* are recorded to be species-rich genera and in hornworts, *Anthoceros* is well represented by species. Pottiaceae, Lejeuniaceae and Notothyladaceae are largely represented in India. The data include total of 340 species as endemic of which 269 species are of mosses, 67 are of liverworts and 4 are of hornworts. Out of 133 rare species 78 are mosses and 53 are liverworts and nearly 14 species of hornworts are recorded as endangered (Dandotya, 2011).

Liverworts

Jungermanniales is the largest group with 62.5 per cent generic representation, Sphaerocarpales (one genus) with 1 per cent and the Calobryales (two genera) with 2 per cent generic representation reported in India. Marchantiales (24 genera) with nearly 23 per cent and Metzgeriales (12 genera) with 11.5 per cent generic representation are the second and third species-rich groups.

Haplomitriaceae: *Haplomitrium* is represented by six species, which are largely confined to Eastern Himalayas (Darjeeling, Jowai, Assam).

* The survey was conducted before the new scheme of classification was proposed, hence the orders and families mentioned are according to previous system of classification.

Sphaerocarpaceae: Only one genus, *Riella* with 2 species is reported from the Varanasi and Central Indian zone in the lake Nal, Ahmedabad.

Metzgeriales: The order is represented by 58 species included in 12 genera and 7 families, richly distributed in Eastern Himalayas. Twenty one species are distributed in Western Himalayas, 9 species are common to South India and 14 to the Eastern Himalayas. Eight species found in the central and north Indian zone.

Marchantiales: Marchantiaceae, Aytoniaceae are richly represented in the Himalayas and Western Ghats. Sauteriaceae with 2 genera is mainly found in the Western Himalayas. Ricciaceae with 2 genera with over 26 species is distributed in plains and hills.

Jungermanniales: Nearly 65 genera and 520 species mainly in the Himalayas and South India of which 108 species are endemic.

Hornworts

Anthocerotales: Nearly 34 species are found in India comprising 5 genera, of which 13 species are confined to the Himalayas, 10 to South India and 1 species to Central India. Nearly 19 taxa are endemic. *Anthoceros* has 9 species of which 5 are endemic. Out of a 12 species of *Folioceros* in India, 9 are endemic. *Phaeoceros* with 4 taxa in India is widely known in the Western Himalayas. *Notothylas* shows maximum representation in the Western Ghats.

Rare and Endangered Taxa in India

The natural habitat of bryophytes in India includes Himalayas, Eastern and Western coasts of peninsular India and hill ranges of central and eastern parts of the country (Dandotiya, 2011; Kumar & Srivastava, 2011). Developmental activities have destroyed the habitat of many species causing a threat to their survival. As a result several species are endangered while others have become rare.

Rare and Endangered Liverworts

Takakia ceratophylla, Haplomitrium blumii, H. hookerii, Anthelia julacea, Ptilidium ciliare, Isotaches indica, Schiffneria levieri, Blasia pussila, Calycularia crispula, Metzeria conjugate, M. furcata, Monoselenium tanerum, Conocephalum supradecompositum, Wiesnerella denudate, Lunularia cruciata, Preissia quadrata, Sauteria alpine, Exormotheca ceylonesis, Ricciocarpus natans, Riella affinis, Apotreubia nana, Geocalyx greveolens, Delavayella serrata, Bazzania assamica, Fossombronia indica, Riccardia villosa, Exormotheca tuberifera, Cyathodium denticulatum, C. indicum, Riccia reticulate and *R. pandei* are rare and endangered genera/species.

Rare and Endangered Mosses

Aloina, Distichium, Hedwigia, Homamallium, Leskeella, Aerobryum, Aliergon, Gammiella, Isothecium, Ortholimnobium, Pleurozium, Leptodontium, Meteoriella,

Wilsoniella, Pottia, Astomum, Microcampylopus, Thiemma, Trachycarpidium, Nanothecium, Pleurozium and *Spharotheciella* are placed in this category.

Response to Global Climate Change

Climate change is now recognized as one of the most significant threats to nature and biodiversity. It is a major driver of ecosystem change and degradation (Hooper *et al.,* 2012). As a consequence of global warming, significant losses in bryophyte diversity can be expected particularly in areas harbouring large number of species, such as boreal forests of higher latitude, alpine biomes and higher altitudes on tropical mountains. This decline of bryophytes is expected to lead to alteration of ecosystem structure, and function, nutrient cycling and carbon balance. Bryophytes are generally drought avoiders, are highly dependent on external water and many species are known to be sensitive to relatively high temperatures. Most bryophytes are capable of surviving short to moderate periods of desiccation (Proctor *et al.,* 2007) and because of their poikilohydric nature, the potential of short-term thermal acclimation of hydrated bryophytes (gametophytes) is very low. This indicates that in hydrated state bryophytes are sensitive to high temperature (Meyer & Santarius, 1998). Many bryophytes show a tendency to be highly sensitive to elevated temperatures. Due to their poikilohydric adaptation bryophytes lose water rapidly when temperature rises and relative humidity drops. Reduced bryophyte cover has been reported due to vascular plant-range shifts induced by higher temperature during the growing season and higher evaporation rates in high latitude ecosystems (Fremstad *et al.,* 2005).

However, the detailed studies on bryophytes and climate change have largely remained inconclusive.

Chapter 2

Phylum – Marchantiophyta – *The Liverworts*

Liverworts for a long time were grouped together with mosses and hornworts in the Division Bryophyta, within which liverworts made up the class **Hepaticae,** now known as *Marchantiopsida. However, since this grouping made Bryophyta appear paraphyletic, which has now been revised to a **monophyletic assemblage (Shaw & Renzaglia, 2004), the liverworts are now given the rank of Phylum. The use of the division name Bryophyta *sensu lato* is still found in the literature, but new concepts of classification refer to Bryophyta in a restricted sense to include only the mosses.

CLASS: **MARCHANTIOPSIDA** with 2 Subclasses: Blasiidae and Marchantiidae

SUBCLASS: **MARCHANTIIDAE** with 4 orders: Sphaerocarpales, Neohodgsoniales, Lunulariales and Marchantiales

ORDER: **MARCHANTIALES** has 15 families: Marchantiaceae, Aytoniaceae, Cleveaceae, Monosoleniaceae, Conocephalaceae, Cyathodiaceae, Exormothecaceae, Corsiniaceae, Monocarpaceae, Oxymitraceae, Ricciaceae, Wiesnerellaceae, Targioniaceae, Monocleaceae, Dumortieraceae.

* Although Marchantiopsida are resolved as monophyletic, traditional relationships among taxa generally are not supported by molecular data. The classical morphological separation of this liverwort class into three orders *i.e.,* Monocleales, Sphaerocarpales and Marchantiales is challenged by nucleotide sequence data (Wheeler, 2000).

** Most commonly bryophytes are viewed as a grade of three monophyletc lineages, with an uncertain branding order (Qui *et al.,* 1996). Recent analyses of amino acid sequence based on entire plastid genome has provided support for a monophyletic bryophyte assemblage; however these results must be viewed with caution because of limitations in taxon sampling (Nishiyama *et al.* in press).

The division Marchantiophyta includes the thallose and leafy liverworts and consists of approximately 8000 species spread over about 350 genera. They possess characters distinct from the mosses.

- The gametophytes are dorsiventral and flat. They may be thalloid (thallose) as in *Riccia* or differentiated into leaves and stem (foliose) as in *Porella*.
- In foliose types, the leaves are arranged in two or three rows on the axis and are always without a midrib.
- Rhizoids are the organs of attachment and are never branched.
- The gametophyte is internally made up of uniform tissue or various types of tissues with many chloroplasts without pyrenoids.
- The sex organs develop from superficial cells on the dorsal surface of the thallus or are terminal in position in leafy liverworts (*Porella*). The sex organs are not associated with paraphyses.
- The sporophyte (spore bearing structure) may be simple, or differentiated into foot and capsule, or into a foot, seta and capsule.
- The archesporium develops from the endothecium of the embryo.
- Spores and elaters arise from the archesporium and elaters when present are generally single celled.
- Spores germinate to form a reduced two or three-celled protonema, more frequently called germ tube.
- The sporophyte is completely dependent on gametophyte for its nutritive supply.
- The wall of sporophyte (capsule region), spore bearing structure is one to several layered thick and lacks stomata.
- The dehiscence of capsule is normally into four longitudinal valves but in some it is irregular generally along weak lines or in irregular fragments and spores are shed simultaneously.
- There is immense diversity in this group of liverworts which stands out from groups of mosses and hornworts.

* *As discussed by many authors (e.g. Wheeler, 2000; Boisselier-Dubayle et al., 2002; Forrest et al., 2006; Long, 2006), there is little congruence between past morphology based classifications of the Marchantiidae (e.g. Bischler, 1998; Crandall-Stotler & Stotler, 2000) and the phylogenetic relationships resolved in recent multi-locus analyses. As a consequence, major modifications in the classification of the subclass have been made (Crandall-Stotler, 2009). The most significant changes include the following; transfer of Neohodgsonia from the Marchantiaceae to its own family and order; the recognition of a monogeneric Lunulariales; the incorporation of Monocleales and Ricciales into the Marchantiales, with the Monocleaceae aligned near the Dumortieraceae and the Ricciaceae aligned close to Wiesnerellaceae; the*

transfer of Peltolepis from Monosoleniaceae to Cleveaceae; and the recognition of a monogeneric Dumortieraceae (Long, 2006). The main evolutionary trend in the Marchantiidae leads to reduction and simplification.

Class: Marchantiopsida

Subclass: Marchantiidae

- Plants are usually thalloid and rarely leafy.
- Thallus is typically differentiated on the dorsal and ventral surface, with well-defined or simple air chambers and pores on the dorsal surface.
- Scales and rhizoids are present on the ventral surface of the gametophyte.
- Antheridia are present in chambers in specific places depending on the species. In some species antheridial/perigonial chambers are found scattered on the dorsal surface of the thallus, while in others, are seen clustered on the main thallus or on a specialized, raised receptacle (antheridiophore).
- The location of the archegonia also varies depending on the species. While in some species they are found on the dorsal surface of the thallus, in others, are seen raised on a stalked receptacle, termed the archegoniophore.
- Even when the sporophyte reaches maturity, the seta is typically short, and may even be lacking in some taxa.
- The capsule wall is typically unistratose and dehiscence is by longitudinal valves or slits.

Order Marchantiales

The order includes liverworts with complex thallus structure found in approximately 15 families with 26 genera and most of them are worldwide in distribution. The characteristic features of the group are:

- The plants are 'complex thalloid liverworts'.
- The thallus is ribbon-like, dichotomously branched and dorsiventral.
- Majority of the taxa possess internally differentiated thallus with photosynthetic and storage zones.
- The air chambers occur on the dorsal side of the thallus; such chambers open outside by defined air pores.
- The ventral portion of the thallus consists of parenchyma which acts as a storage tissue; oil and mucilage cells may be present in this region.
- The scales and rhizoids are present on the ventral side of the thallus.
- The rhizoids are of two types (smooth-walled and tuberculate) and help in anchorage and absorption in water.
- The unistratose scales are arranged in one or more rows and are involved in water retention and conduction.

- The antheridia and archegonia may be found directly on the dorsal surface (imbedded) of the thallus along the midrib or they may be present on the special branches known as antheridiophores and archegoniophores respectively.
- In most cases, the capsules possess single layered jacket or wall (unistratose) and they open by two longitudinal sutures but never by four valves.
- The elaters with spiral thickenings may or may not be present. They have a role in spore dispersal.
- The sporophyte may be simple without seta and foot as in *Riccia* or it may be differentiated into foot, seta and capsule as in *Marchantia*.

** **ORDER: MARCHANTIALES** Limpr. in Cohn, Krypt.-Fl. Schlesien 1:239, 336 (1877). Thallus usually differentiated; epidermis with either simple or compound air pores (rarely absent); ventral scales in 2 to 10 rows, sometimes absent, usually with 1 to 3 (to 6) appendages; rhizoids usually dimorphic, sometimes only smooth; idioblastic oil cells usually present; monoecious or dioecious; perigonial chamber positions variable; sporophyte positions variable; involucres bivalved, cup-shaped, scale-or flap-like, or tubular, sometimes absent; pseudoperianths absent or present; seta usually short or absent; elaters usually present; capsule dehiscence by longitudinal valves, longitudinal slit or lid, sometimes cleistocarpous; gemmae present in a few taxa.*

Box 2.1

Within this monophylogenetic assemblage, are several morphologically isolated elements that represent products of deep divergences (Renzaglia *et al.*, 2000).

Since the starting point of liverwort nomenclature (Linnaeus, 1753) and the beginning of their systematic treatment (Endlicher, 1841), hepatics have been organized into three groups based on growth forms: 1) complex thalloids, 2) simple thalloids and 3) leafy liverworts.

Morphological studies supported the concept that simple thalloid liverworts are more closely related to leafy types than to complex thalloids.

Classification schemes reflect this interpretation that hepatics may be divided into two groups: Marchantioid or complex thalloid liverworts and Jungermannioid liverworts including leafy and simple thalloid forms.

The group has oil bodies which accumulate terpenoids. It has external appendages to protect fragile tissues *e.g.*, mucilage papillae, hairs, scales, bracts. Superficial sex organs are protected by flaps of tissues, leaf lobes, young leaves or modified branches.

The sporophytes reach maturity within confines of protective gametophytic tissue *e.g.*, perigynium and calyptra. Additional protection is provided by perianth, pseudoperianth and bracts. Sporophyte is completely dependent on gametophyte. Majority of liverwort sporophytes are differentiated into foot, seta and capsule and in others seta and/or foot may remain vestigial or absent. Sterile and elongated hygroscopic elaters help in spore dispersal.

Family: Marchantiaceae Lindl.

- Marchantiaceae Lindl., Nat. Syst. Bot., ed. 2:412 (1836).
- Thallus differentiated into dorsal and ventral surfaces, with compound air pores on dorsal surface; ventral scales in 2 to 10 rows, with 1 to 3 appendages.
- Perigonial/antheridial chambers with an antheridia, aggregated on stalked receptacles, with the receptacles bearing a disc with compound pores and the stalk with 2 to 4 rhizoid furrows.
- Sporophytes on stalked receptacles, with, receptacles bearing compound air pores on the disc and the stalk with 2 to 4 rhizoid furrows.
- Involucre bivalved or cup-shaped.
- Pseudoperianth present, campanulate; seta remaining short initially; Capsule dehiscence is by irregular valves.
- Gemmae absent, or present in cup-shaped receptacles, gemma cups *(Marchantia), Bucegia, Radian, Marchantia L., Preissia, Corda.*

Marchantia

Habitat and Distribution

It grows on a wide variety of habitats, including cliffs, closed forests, alpine heathlands, peat bogs, minerotropic fens, springs, swamps, grasslands, and tundra. It is most often commonly found on moist or wet mineral soil, especially in recently burnt areas (Durand, 1908). Common liverwort grows best in subcalcareous soil conditions (pH 6.0), under full sunlight (Gilley, 1982). About 65 species are distributed world over.

Gametophyte

There are two schools of thought to interpret the simplicity or complexity of structures in *Marchantia.* One school believes that *Marchantia* is more advanced when compared to *Riccia* while the other believes that *Marchantia* thallus got simplified to give rise to *Riccia*. The following is an account on which the clarification can be sought.

The haploid spore after falling on the ground under favourable conditions germinates to produce a germ tube which represents the reduced protonema. Later, an apical cell with two cutting faces gets organized. The flat thallus is produced after the apical cell cuts cells on its cutting phases. It must be remembered that out of a tetrad of spores, two spores give rise to the male and two to the female thallii. The thallus is anatomically much complex and bears rhizoids and scales of atleast two types each on the ventral surface. It has two distinct zones, storage and assimilatory, where the latter has photosynthetic filaments inside air chambers that open outside by distinct air pores.

Although *Riccia* appears to be simple if not primitive, but the breeding experiments have indicated that evolutionary predecessors of *Marchantia* have given rise to other genera through simplification. If this line of thought is correct then *Riccia* being a reduced form, is more advanced. On the other hand, both *Marchantia* and *Riccia* may be considered as evolved forms. Though the stalked reproductive structures elevate seta-less capsule and help spore dispersal, in *Riccia* spores are long lived. Despite its rudimentary structure, the *Riccia* sporophyte is also a well-adapted structure (Bell & Hemsley, 2000).

Morphology

- Thallus is flat dorsiventral, plagiotropic, and dichotomously branched with a notched apex.
- Apical notch is distinct with a median groove on the dorsal surface and a corresponding ridge on the ventral surface.
- The dorsal surface under the hand lens shows rhomboidal areas with a small black spot in the centre. The rhomboidal areas correspond to the air chambers present inside the thallus and the black spots correspond to the air pore **(Fig. 2.1).**
- The ventral surface bears two to four rows of scales, where the outermost extends beyond thallus margins. There are two types of rhizoids arising along the ridge.
- The thallus with vegetative growth shows gemma cups and those with sexual reproduction have stalked discs or receptacles.

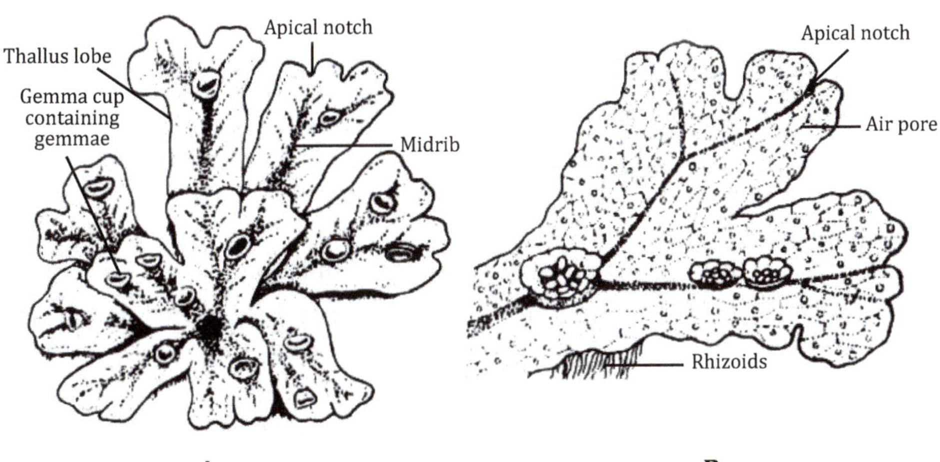

Figure 2.1*A&B Marchantia. A* - Shows a dichotomously branched thallus with apical notch and median groove or midrib. Also seen are gemma cups. *B* - Dorsal surface shows dot-like structures that represent air pores. Gemmae are seen inside the gemma cups.

Scales

- Ventral surface bears scales along the midrib where they are arranged in two to four rows on either side of the midrib as in *M. polymorpha.*
- Median scales are pale mauve, present in one row on either side of midrib. These scales narrow upwards, and are deeply constricted where they are joined with a broad rounded appendage, brown or tinged with purple **(Fig. 2.2A).**

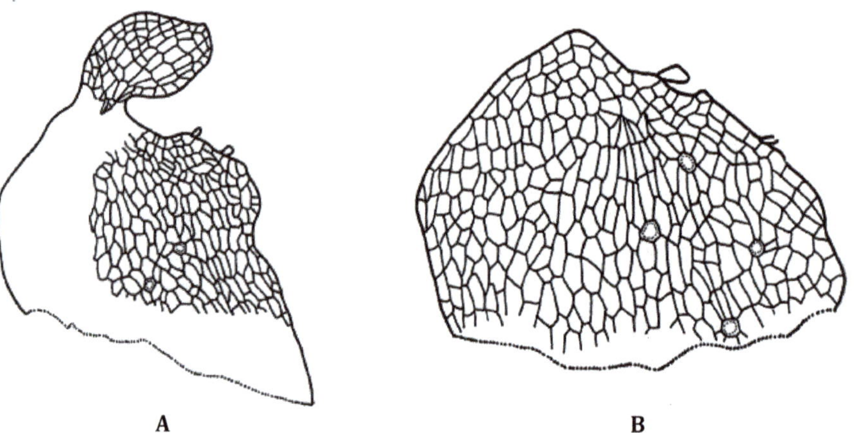

Figure 2.2*A&B Marchantia* Whole mounts. *A* - Median ventral scale with appendage. *B* - Laminar ventral scale.

- Luminal or intermediate scales are large, hyaline or pale mauve, in one row on either side of and lateral to median scales, almost wedge-shaped, but with apex rounded, base flatly arched **(Fig. 2.2B).**
- Marginal scales are the largest, sharply toothed, crenulate (irregular with shallow notches) or almost entire and usually project beyond thallus margins. They are hyaline or brownish, oblong or ovate, with rounded apex (Perold, 1995).
- The rhizoid initials in the scales give rise to rhizoids.
- Mc Conaha (1941) noted that scales were passively wetted and formed along with rhizoids, a capillary network that acted as a conducting system for the plant.
- The scales also are believed to help orienting the pegged rhizoids.
- They store/hold some amount of water which helps to keep the thallus moist.

Rhizoids

- Two types of rhizoids (similar to *Riccia*) occur on the ventral surface along with the scales and make an efficient capillary system.
- Tuberculate rhizoids individually originate from the lower superficial cells of the apical meristem. Pegs exist in the lumen of rhizoids, which always lie parallel to the thallus surface and converge toward the midrib.
- Smooth-walled rhizoids always exist in clusters in free portions near the midrib in the ventral surface of the thallus. These smooth rhizoids always grow toward moist soil and lie perpendicular to the thallus surface (Cao *et al.,* 2014).

Anatomy

- The gametophytic thallus is complex in its organization.
- It has two basic zones - storage and assimilatory, typical of Marchantiales **(Fig. 2.3).**

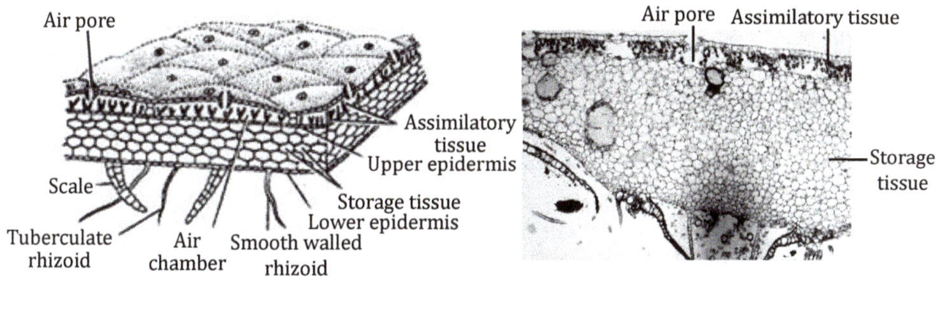

Figure 2.3*A&B Marchantia* V.S. Thallus. *A* - Line diagram to show internal 3D construction of the thallus with two distinct regions, storage and photosynthetic. Also seen is the upper epidermis with air pores. *B* - Stained preparation of the same.

- The assimilatory region is simple parenchymatous, thin walled and compactly arranged. The cells contain a few chloroplasts and abundant starch.
- The lowermost cells of storage zone form lower epidermis from which arise scales and rhizoids.
- The assimilatory or the photosynthetic region is complex with outermost layer, the upper epidermis which is interrupted by barrel-shaped air pores.
- The air pore is an intracelluar space circularly surrounded by four tiers of cells forming a barrel shape. Each tier surrounding the air pore consists of five or six cells in height, and the lowermost cells form the cruciform slit of the pore.

- The pores lead to the air chambers which are separated from each other by single layered cells, four or five celled in height.
- From the floor of the air chambers arise, short, photosynthetic filaments which are branched and loosely arranged **(Fig. 2.4; Plate I.1).**

Do You Know?

Most species of liverworts grow in association with arbuscular mycorrhizal fungi, and symbiotic fungi typically penetrate the plant body through the rhizoids. Such fungi were found to be present in *Marchantia polymorpha* subsp *montivagans* growing in natural habitats but other subspecies, growing in nutrient-rich habitat did not show this association (Ligrone *et al.*, 2007).

** The construction of the thallus may be equated to a room. There is a floor from which arise chlorophyllose filaments; the pillars are the partition walls and there is a roof with a chimney or the pores. The air pore of Marchantia seems to provide a balance between gas exchange and water vapour loss as efficiently as stomata in sporophytic tissue of other land plants. The architecture and hydrophobicity of the pore seem to prevent the penetration of liquid from outside into the intercellular space of air chambers (Schonherr& Ziegler, 1975). Though air pore does not exhibit an opening and closing movement like stomata but, cells surrounding the bottom of the pore change shape to adjust width of the opening of the slit (Walker & Pennington, 1939). The air chamber equipped with air pores and assimilatory filaments is considered to facilitate gas exchange for photosynthesis, and help in transpiration and respiration (Meyer et al., 2008). The fact that gametophytes of land plants did not develop any pores and stomata except for complex thalloid liverwort and hornworts might be interpreted as as evidence that the gametophyte is phylogenetically the water generation and the sporophyte is the newly acquired aerial generation (Ziegler, 1987).*

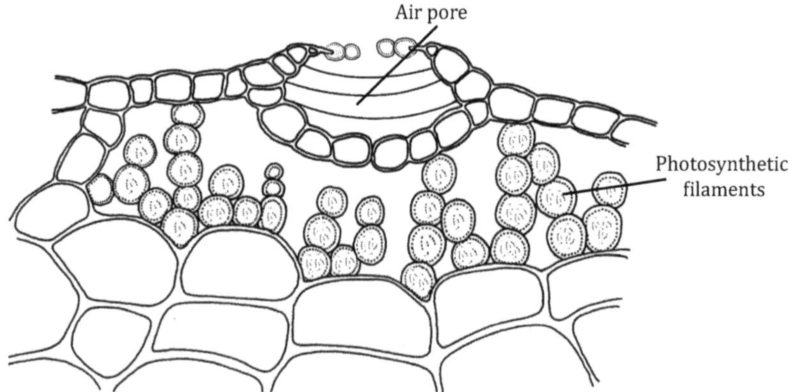

Figure 2.4 *Marchantia* V.S. Thallus, a diagrammatic sketch to show details of the photosynthetic region.

Reproduction

Vegetative Propagation

It occurs by progressive death and decay of the thallus where decay starts at the posterior end and progresses towards apex. Adventitious branches which arise either from ventral surface of thallus or the archegoniophore also give rise to new plants. However, the most frequent form of vegetative reproduction is by gemmae produced in gemma cups or cupules formed on the dorsal surface.

- Gemma cups which are raised structures on the dorsal surface, have a frilled margin and a base or floor from which arise gemmae **(Plate I.2).**
- The internal organization of the cups with assimilatory and storage zone appears similar to vegetative thallus **(Fig. 2.5A).**
- The gemma is attached to the floor of the cup with a one celled stalk **(Plate I.3).**
- Mature gemma is a discoid multicellular body, thick in the centre and thin towards the margin.
- Two notches, one each on the lateral margin harbour apical cells **(Plate I.4).**
- The cells forming the discoid body are of various types - rhizoidal, chlorophyllose and cells that contain oil **(Fig. 2.5B)**.
- A few cells at the floor develop into hairs with mucilaginous activity.
- When water is available, the hairs secrete mucilage and swell upon absorbing moisture, thus exerting pressure on the stalk of the gemma. The gemma breaks at this point and is carried by water or rain or are splashed out (Plate VIII.1).
- Gemmae have been observed to splash as far as 120cm away from the parent.
- The gemmae have bilateral symmetry but once they fall on the surface after dissemination, they show dorsiventrality and get anchored to the substratum by rhizoids arising from rhizoidal cells and this side of thallus differentiates as ventral surface.
- The two apical cells of the gemmae placed diagonally opposite to each other become active and start dividing.
- The cells in the centre start decaying and two different thalli by apical cell activity are formed when decay reaches apical cell.
- Gemmae from male gametophyte generally form male thalli and those from the female form female thalli.

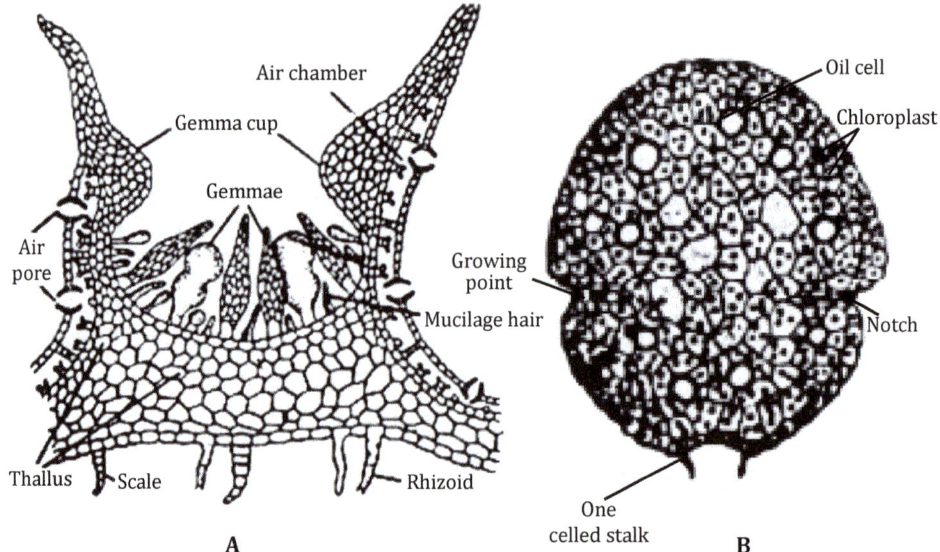

Figure 2.5 *A&B Marchantia* V.S. Thallus through gemma cup. *A* - Basic organization of the cup is similar to the vegetative thallus and shows a photosynthetic and assimilatory zone. The gemmae are attached to the floor of the cup. *B* - Each gemma shows a small stalk, growing points, and rhizoidal cells.

Sexual Reproduction

During sexual reproduction, the stalked structures referred to as receptacles or archegonio- and antheridiophore bearing archegonia and antheridia or carpocephala, respectively arise from the dorsal surface. Both, the stalks and the spread out eight lobed discs are similar to thallus in construction with storage and assimilatory zones. The latter bear air chambers, barrel-shaped air pores and photosynthetic filaments as in vegetative thallus. In male plant, antheridia lie within the chambers formed by the overgrowth of disc tissue around the antheridia **(Plate I.5)**.

Antheridiophore

- The antheridiophore is a stalked receptacle, bearing male sex organs known as antheridia.
- The antheridiophore has 1 to 3cm long stalk that bears a slightly convex usually eight lobed peltate disc at its apex.
- Each lobe of the peltate disc has a growing point at the tip and this represents the apex of a dichotomous branch.
- The internal structure of the peltate disc is similar to that of the thallus.
- The upper epidermis is interrupted by a number of barrel-shaped air pores, each opening below into an air chamber. The air chambers have many simple or branched photosynthetic filaments.

- The antheridial chambers alternate with the air chambers.
- The margin of antheridial receptacle is wraped upward, creating a pool of water.
- Each antheridial chamber contains a single antheridium and opens externally by a pore, occasionally referred to as the ostiole **(Plate I.6).**
- On each lobe of the disc, antheridia arise in acropetal succession, *i.e.,* the oldest near the centre and the youngest towards the apex of the lobe.
- Antheridium has a small stalk and a globular body with a single layered jacket that encloses biflagellate antherozoids at maturity **(Fig. 2.6).**

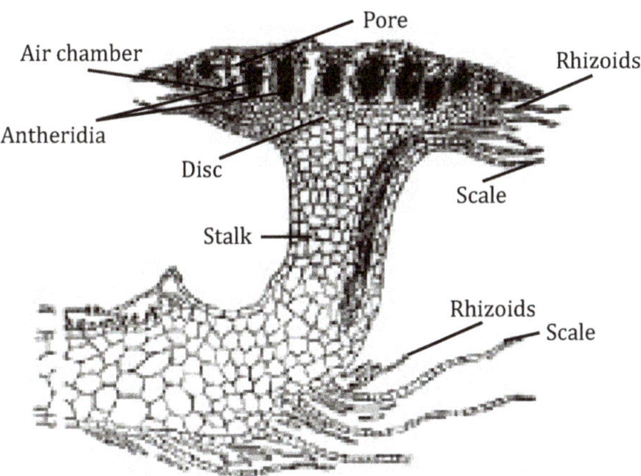

Figure 2.6 *Marchantia*. L.S. Antheridiophore showing a disc with antheridial chambers. There arises one antheridium per chamber. The chambers open to outside by pores. The stalk is similar to thallus in construction.

Archegoniophore

- It has a stalk and a lobed disc where the stalk is short in developing stages **(Plate I.7).**
- The disc is convex and is dichotomously branched. The internal structure is similar to that of thallus. Outermost is the epidermis interrupted by air pores. These open into air chambers with branched photosynthetic filaments.
- At maturity disc becomes eight lobed due to the dichotomies.
- In some species digitate rays are seen to arise along the lobes that retain water droplets.
- Each lobe bears a notch and behind this apex, is a group of acropetally arranged radial row of archegonia with youngest towards the notch.
- The stalk is prismatic with definite ridges and furrows.

- As the first archegonium is fertilized, there is upward growth of the stalk in the centre of the disc that results in inversion of marginal portion of disc that carries archegonia **(Plate I.8).** As a result, the archegonial necks are directed towards the ground **(Plate I.8 inset)** and the position of archegonia is reversed – the oldest archegonium is towards periphery of the disc and the youngest near the stalk **(Fig. 2.7A&B).**
- Fertilization of rest of the archegonia occurs later as they hang down inverted.
- The archegonium is flask-shaped with venter, the swollen portion containing egg and ventral canal cell and a long neck formed of six vertical rows and has four to eight neck canal cells **(Fig. 2.7B).**
- Cover cells of the neck are not much distinct.

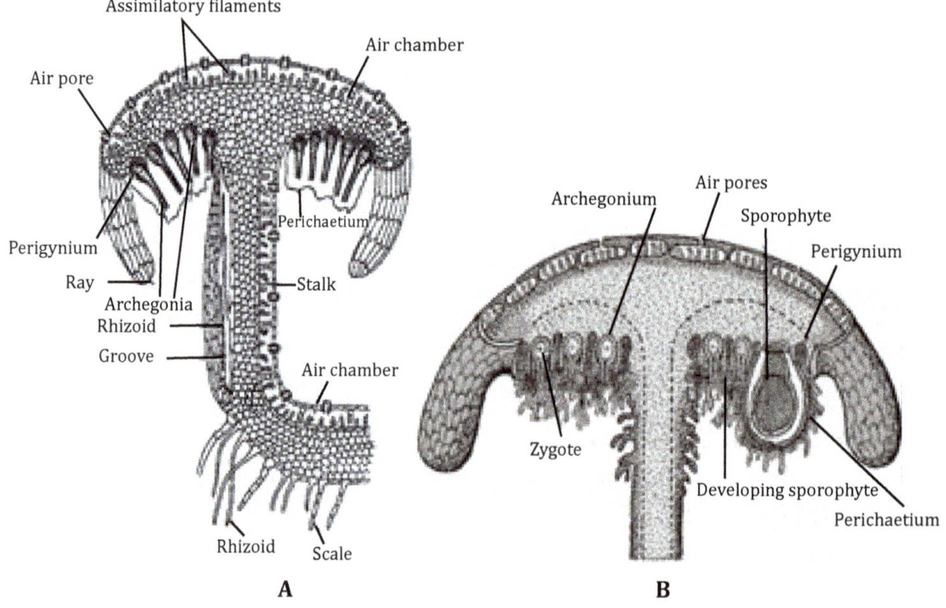

Figure 2.7*A&B* *Marchantia* L.S. Archegoniophore. *A* - The disc is inverted and carries the youngest archegonium towards the stalk while the oldest is towards the periphery of the disc. *B* - Archegonia at various stages of development. Also seen is a young sporophyte. The three protective layers that cover all the archegonia of the lobe are distinct.

Fertilization and Post-fertilization Changes

Fertilization is brought about when antherozoids reach the archegonia through splash cup mechanism (discussed in last chapter). Fertilization mediated by water leads to zygote formation, the first stage in sporophytic generation. This phase is more elaborate than in *Riccia* and is marked by post-fertilization changes in the disc that bears several fertilized or mature archegonia. With the first archegonium getting fertilized, the stalk of archegoniophore elongates and the disc shows meristematic activity in the central region. Due to these changes, the disc becomes inverted and the archegonia now hang with their necks down. As the disc gets inverted, the youngest archegonium is placed next to the stalk and the subsequent fertilization is carried out as the water carrying antherozoids trickles down the disc along its sides. After the inversion, an involucral sheath or perichaetium around a group of archegonia of a lobe is formed from the disc along its side. The stalk of the archegonium also elongates and the venter wall divides periclinally forming two layered calyptra which is a protective structure around the zygote. Another tissue, perigynium or pseudoperianth arises from the base of the venter which eventually encloses the sporophyte. Thus, post-fertilization, three protective sheaths are formed - from outside inwards are perichaetium, perigynium and calyptra. All these structures are haploid.

Sporophyte

The zygote undergoes first division and forms an upper epibasal cell and lower hypobasal cell. In subsequent divisions, tissues of capsule are derived from epibasal cell while seta and foot are derived from hypobasal cell. After a few divisions, a multicellular embryo with two embryonic tissues - amphithecium and endothecium is formed. The amphithecium gives rise to one layered jacket, a diploid structure and endothecium forms the archesporium or the fertile tissue which gives rise to sporogenous tissue. In *Marchantia* all the sporogenous cells do not add to fertility but some may form sterile structures as well. As in *M. polymorpha* only about half of the sporogenous cells form spore mother cells. Each sporogenous cell divides to form thirty two Spore Mother Cells (SMC) and each SMC undergoes meiosis to form a spore tetrad from which ultimately for haploid spores are produced, thus establishing a gametophytic phase. Thus, each sporogenous cell will produce 128 spores. The sporogenous cells which add to sterility become elater mother cells and form elaters (one elater mother cell forms one elater). The elaters are hygroscopic in nature and help in dispersal of spores. In some species a sterile apical cap-like structure is also present. Therfore, if one sporogenous cell is sacrificed in elater formation, about 128 spores are sacrificed.

The mature sporophyte has a short seta which elongates once the spores are ready for dispersal. When seta elongates, the capsule bearing spores and elaters is pushed out of the protective coverings. Depending upon the conducive environmental conditions of moisture and temperature, the capsule which is an elongated structure splits longitudinally and the spores are dispersed, facilitated by the hygroscopic movement of elaters **(Fig. 2.8).**

The foot of the growing sporophyte and the adjacent gametophyte tissue form a nutrient-transfer zone, the 'placenta'. In the placenta region, specialized transfer cells develop which have wall ingrowths (Ligrone *et al.*, 1993).

The number of spores in a single capsule is estimated as 300,000 and the total spore output for a single archegoniophore is > 7,000,000 if an archegoniophore produces 24 sporophytes (O'Hanlon, 1926).

Salient features of Sporophyte

Sporophyte develops inside the archegonium.

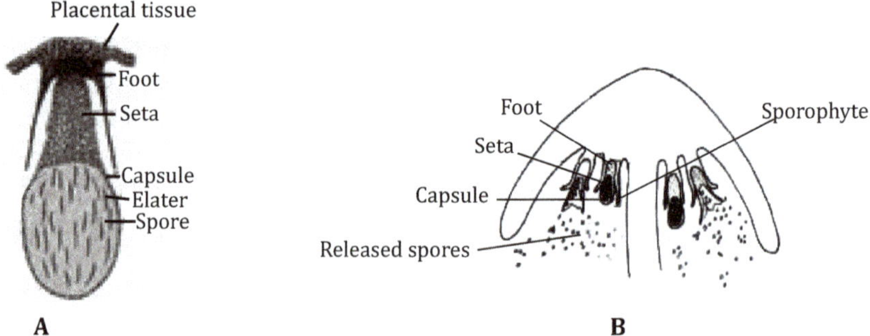

Figure 2.8*A&B Marchantia. A* - Mature sporophyte. *B* - Line diagram to show spore release at maturity. Note the inverted sporophyte. (Source: http://lungtp.com/plantae/e_defce.html). (Source: https://www.anbg.gov.au/bryophyte/reproduction-dispersal.html).

- Since a group of archegonia of a lobe is covered by a common involucre or perichaetium, multiple sporophytes at different stages of development are seen in a single involucre **(Plate I.9).**
- The sporophyte, an inverted structure is enclosed by three coverings from inside out: Calyptra, Perigynium (or pseudoperianth) and Perichaetium (or involurce).
- It is differentiated into a foot, seta and capsule **(Plate I.10A).**
- Foot is bulbous and remains embedded inside the gametophyte. Its function is to anchor the sporophyte and draw nutrients from the gametophyte **(Plate I.10B).**
- Seta is middle portion and short in the young sporophyte. It plays an important role in capsule dehiscence.
- The capsule is elongate with a single layered jacket, inside which lie many spores and elaters **(Plate I.11).**
- When the spores are mature, the seta elongates pushing the capsule through the protective layers, perichaetium and perigynium **(Plate I.12).**
- The jacket then responds to the outside conditions of moisture and splits into various lobes from apex downwards exposing the elaters and the entangled spores.

- Elaters are spindle-shaped and each possesses spiral thickening bands **(Plate VIII.6A-C)**. These are hygroscopic and undergo twisting movements that help in the dispersal of spores.
- The spores germinate immediately after they are shed, forming a slender germ tube. A cell at the tip is cut off which acts as an apical cell from which thallus is produced.

The life cycle of the liverwort is depicted in **Figure 2.9**.

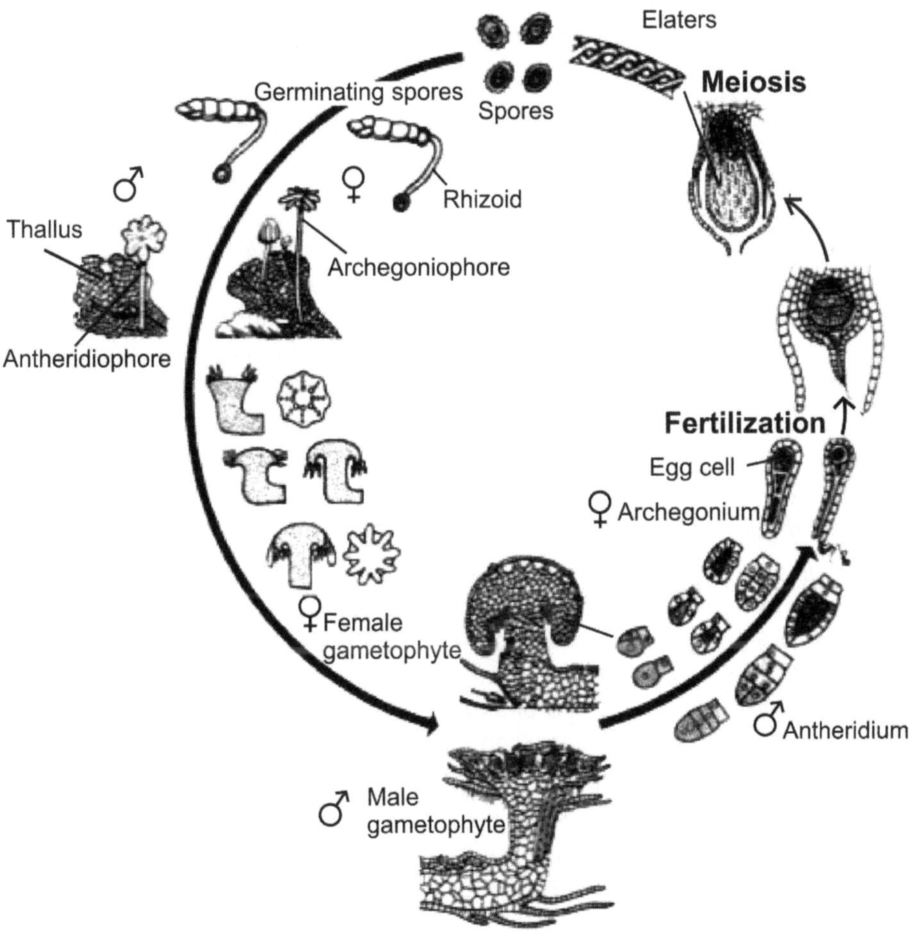

Figure 2.9 ***Marchantia*** **Life cycle showing the haploid dominant gametophytic phase alternates with diploid dependent sporophytic phase.**

Do You Know?

Marchantia has a relatively small genome size of 280 Mbp distributed among eight autosomes plus one sex chromosome (Okada *et al.*, 2000), making it a worthy model organism amenable to genomics-style investigations.

Check Your Knowledge

1. How will you identify *Marchantia* thallus in field?
2. The gametophyte of *Marchantia* is a complex structure. Comment.
3. Explain the water absorbing and retention structures in *Marchantia*.
4. Though there is a great sterility in the sporophyte, yet *Marchantia* is known to be a good colonizer. Justify.
5. Describe asexual reproduction in *Marchantia*.

**Whether Marchantia exhibits evolutionary slow molecular rate compared to Riccia has been discussed by Villarreal et al. (2015). The work was based on 1) presence/absence of carpocephala or receptacles; (2) presence/absence of pegged rhizoids;(3) presence/absence of thallus epidermal pores; (4) presence/absence of air chambers; and (5) presence/absence of photosynthetic filaments.*

The study proved that the phylogenetic position of Marchantia is firmly supported in the earliest divergent lineage of the Marchantiales. The evidence is that compound air pores occurring on both thalli and carpocephala (as in Marchantia)is ancestral in complex thalloids, with various types of simple pores derived.

They further argued that genera like Riccia and Cyathodium have faster rate of evolution in both organellar and nuclear markers. Both are annuals and are short-lived. A correlation between a faster rate of molecular evolution and shorter generation time has been demonstrated in herbaceous angiosperms (Smith & Donoghue, 2008) and mammals (Bromham, 2009).These taxa have reduced or simple sporophytes, and very large number spores that are released when the sporophyte and surrounding thallus disintegrate at the end of the growing season (Bischler, 1998).

Though some authors (e.g., Leitgeb, 1881) have argued that the lack of carpocephala is ancestral in Marchantiidae, but the Villarreal et al. (2015) strongly support Goebel's hypothesis (Goebel, 1930) of secondary carpocephalum losses.

Ricciaceae

- The gametophyte grows in rosette-like habit, formed due to frequent dichotomies of branched thallus.
- The dorsal portion consists of chlorophyllous filaments which enclose air canals or air chambers. The definite air pores are absent though simple pores may be present.

- The ventral portion of the thallus is parenchymatous and acts as a storage tissue.
- Scales present on ventral surface are generally without appendages being simple in nature.
- The sex organs (antheridia and archegonia) are found along the longitudinal groove on the dorsal side. The arrangement is acropetal with youngest near the growing apex and the oldest is towards the base.
- The sex organs occupy cavities on the dorsal surface.
- Sporophytes lie embedded singly in the thallus; additional protective layers around sporophyte are absent.
- The sporophyte, is a simple structure, a capsule embedded in the thallus; foot and seta are absent.
- Archesporium is represented by spores only, elaters being absent.
- No specific mode of dehiscence is seen as the capsule simply disintegrates at maturity (cleistocarpous).

Riccia

Habitat and Distribution

Riccia is the most species-rich genus within the order Marchantiales with more than 150 species in the world (Jovet-Ast, 2005). The genus first discovered by F. F. Ricci, forms rosettes in the moist land. Common species – *Riccia discolor, R.glauca, R. robusta* and *R. crystallina* are terrestrial while *R. reticulata* is xerophytic (desert species) and *R. fluitans* and *Ricciocarpus natans* are aquatic.

Gametophyte

Plants show two distinct phases, the gametophytic and the sporophytic. The gametophyte is a complex structure on which is dependent the simple sporophyte. According to Bower, Campbell and Cavers, the simplest sporophyte of *Riccia* is the most primitive amongst liverworts.

The gametophytic phase begins with the germination of spore (haploid). The spore whose wall ornamentation is a taxonomic feature, germinates to produce a short-lived highly reduced protonemal phase, also haploid. This gives rise to the flat, dorsiventral dichotomous thallus or the gametophyte which remains attached to the substratum **(Fig. 2.10A)** with the help of rhizoids present on ventral surface. On the ventral surface are also present scales that help protect the thallus from dessication.

Morphology

- Plant body is prostrate, flat, somewhat fleshy and dichotomously branched.
- Due to repeated branching a distinct rosette is seen in *Riccia* **(Fig. 2.10, Plate II.1).**

- Each branch of thallus has a midrib represented by dorsal median groove and the ventral median ridge. It terminates in an apical notch in which resides the apical cell.
- The dorsal surface appears faint green due to chlorophyll in the photosynthetic zone.
- Ventral surface bears numerous rhizoids and scales.
- Scales are arranged in a transverse row, are conspicuous multicellular, membranous, pink to red violet and one cell thick structures.
- Scales form continuous transverse strips towards apex and project forward to protect growing points.
- In aquatic species *R. fluitans,* rhizoids and scales are absent.

Figure 2.10*A&B Riccia* Habit. *A* - Plant growing in nature. *B* - Thallus shows frequent dichotomies to form a rosette. Note the apical notch and the distinct midrib.

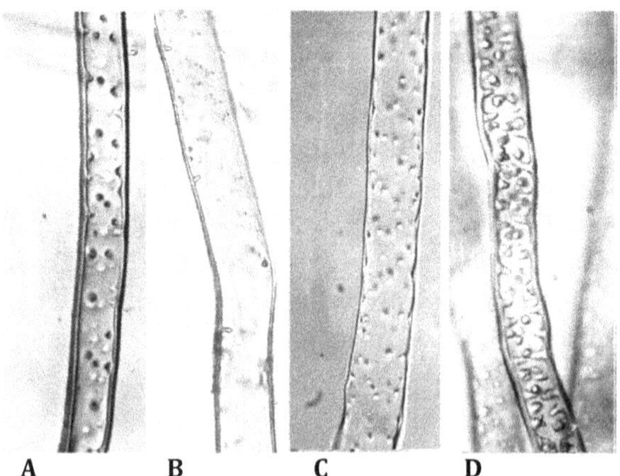

Figure 2.11*A-D* Marchantiopsida. Photographs showing various degree of pegs in the rhizoids. They may be sparse as in *(A)* and *(B)* or they may be dense as in *(C)* and *(D)*.

Rhizoids

- The thallus on its ventral surface bears rhizoids in the midrib region **(Plate II.2)**.
- The rhizoids which are simply hair-like tubular outgrowths from lower epidermal cell, are unicellular and unbranched.
- They appear hyaline due to the absence of chloroplasts but have a more vacuolated cytoplasm.
- Rhizoids are of two types - smooth walled and tuberculate or pegged **(Fig. 2.11A-D)**.
- The smooth walled rhizoids have a two walled structure where both the walls are smooth. These are basically involved in anchorage and absorption.
- The tuberculate rhizoids are also two walled structures but the inner wall grows into the cell forming peg-like structure so as to increase the surface area for absorption of minerals **(Fig. 2.12)**.

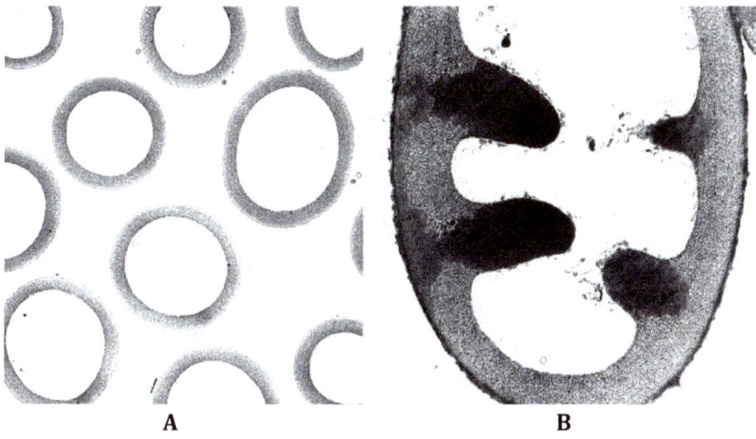

Figure 2.12*A&B* Marchantiopsida. Transmission Electron Micrograph showing smooth rhizoids in *A* and tuberculate rhizoids in *B*.

Scales

- Only simple scales are seen in this genus.
- The scales arise along the midrib and overlap, holding some amount of moisture.
- However, due to frequent dichotomy of the thallus, the scales seem to be present along the margins **(Fig. 2.13)**.
- The scales are membranous pink to violet due to anthocyanins and make the patches look pink in nature.

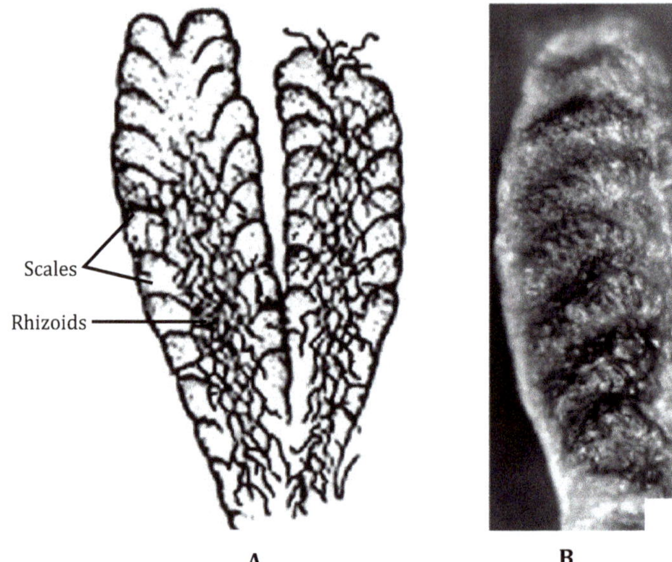

Figure. 2.13*A&B Riccia* Ventral surface. *A* - The scales protective in function, are arranged in an overlapping fashion. The arrangement also helps them store some amount of water that checks desiccation. *B* - Though scales arise near midrib but at maturity, they appear to be placed along the margins.

Anatomy

- Outline of the section shows an apical notch with a narrow wing on either side of the notch.
- The thallus is differentiated into two zones – the photosynthetic or the assimilatory zone and the storage zone **(Fig. 2.14A).**
- Unbranched four to five celled photosynthetic filaments are seen in the green zone **(Plate II.3)**. Each filament is terminated by a cell slightly larger than the rest and is comparable to the epidermal cell, though epidermis is ill-defined.
- The filaments are separated by narrow air canals meant to facilitate exchange of gases **(Fig. 2.14B).**
- The storage zone is formed of compact parenchymatous cells which store starch.
- The lowermost layer of the storage zone is modified into lower epidermis and is represented by small rectangular cells which are more or less vacuolated.
- The region corresponding to the midrib shows rhizoids and scales which arise from lower epidermis **(Fig. 2.14B).**

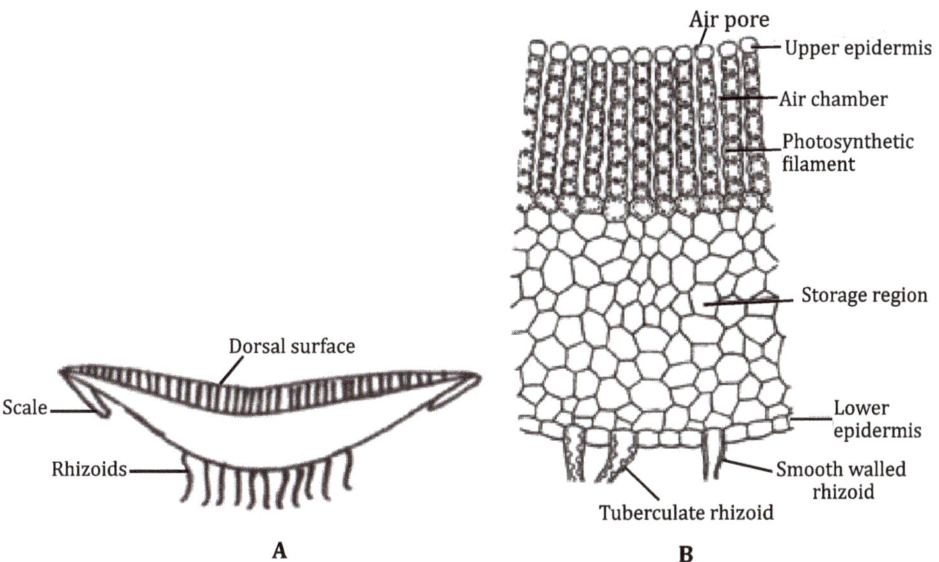

Figure 2.14 ***A&B Riccia*** **V.S. Thallus,** ***A*****-Line diagram.** ***B*** **- Cellular details of the photosynthetic and storage region. The terminal cell of photosynthetic filaments is larger and contributes to the ill-defined epidermis.**

Reproduction

Vegetative Propagation

1. Death and decay of the older parts: Posterior part of the thallus decays and proceeds towards the dichotomy. When the decay reaches dichotomy the terminal ends of the branches separate and grow into new plants.
2. By cell division or gemma formation: The cell divisions occur in young rhizoids developing into gemma-like structure. These structures give rise to haploid plants.
3. By adventitious branches: It occurs in several species of *Riccia* where adventitious branches are produced on the ventral surface of the thallus. These branches detach from the thallus and develop into new gametophytes.
4. By thick apices: It occurs in some species like *R. himalayensis* when the apex of the thallus grows downward into the soil at the end of the growing season. This thick apex survives in the soil developing into a new plant in favourable conditions.

5. By tubers: Some species of *Riccia* as *R. discolor* under unfavourable conditions develop vegetative structures called tubers. The tubers are developed at the apices of these branches. These tubers develop into new plants in favourable conditions.

Sexual Reproduction

After the vegetative growth, plant enters reproductive phase with thalli bearing antheridia and archegonia in cavities on the dorsal surface. Both the sex organs may be borne on the same thallus and a monoecious condition exists or the thalli may be dioecious with separate male and female sex organs. Some species of *Riccia,* like *R. crystallina, R. gangetica, R. billardieri* and *R. glauca* are monoecious or homothallic (*i.e.,* both antheridia and archegonia develop on the same thallus) while other species like *R. curtisii, R. perssonii, R. bischoffii, R. frostii, R. discolor* are dioecious or heterothallic (*i.e.,* antheridia and archegonia develop on different thalli). In hetrothallic *R. curtisii,* there is genotypic determination of sex at time of spore formation, two spores of tetrad develop into female and two into male plants.

Antheridia

- The antheridia arise on the dorsal surface of the thallus and become deep seated in pit-like structures at maturity **(Fig. 2.15A).**
- They are borne along the midrib in an acropetal manner where the youngest is towards the notch.
- Generally one antheridium per chamber is seen.
- An oval antheridium with a short stalk has a single layered jacket, the protective structure **(Fig. 2.15B).**
- It encloses androgonial cells that give rise to androcyte mother cells. Each mother cell forms two androcytes, which finally give rise to antherozoids, the flagellate cells.
- Dehiscence of antheridium is facilitated by presence of water and upon dehiscence, the released antherozoids swim in the water film present in the midrib groove.

Archegonia

- Same as the antheridium – dorsally placed deep pit-like structure harbours a flask-shaped archegonium **(Fig. 2.16A).**
- It has a neck and a venter; where the neck has six vertical rows, four neck canal cells and the swollen venter has a ventral canal nucleus and an egg cell **(Fig. 2.16B, Plate II.4).**
- The entire structure opens outside by means of four cover cells of the neck that protrude out.

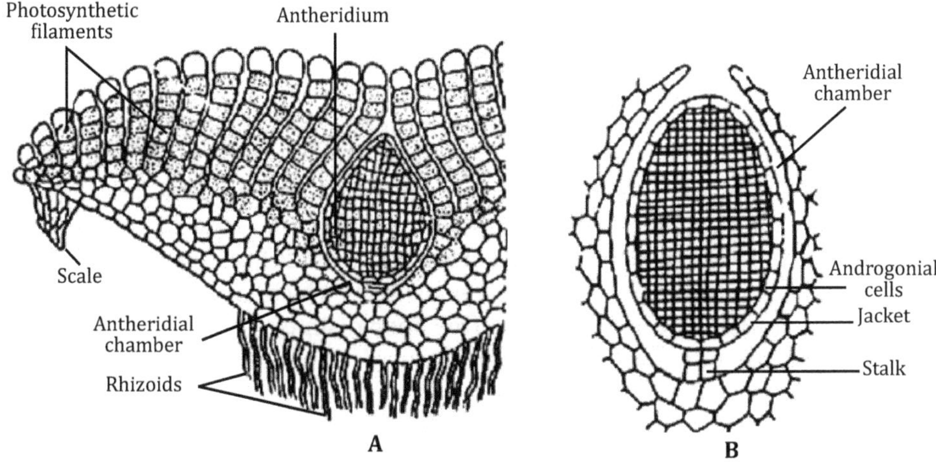

Figure 2.15*A&B Riccia* V.S thallus through antheridium. *A* - The male sex organ is deep seated in the dorsal surface of the thallus. *B* - The antheridium at maturity is short stalked with a globose body, the single layered jacket encloses androgonial cells.

- Fertilization is water dependent event where the archegonial cover cells are removed and neck canal cells and ventral canal cell disintegrate

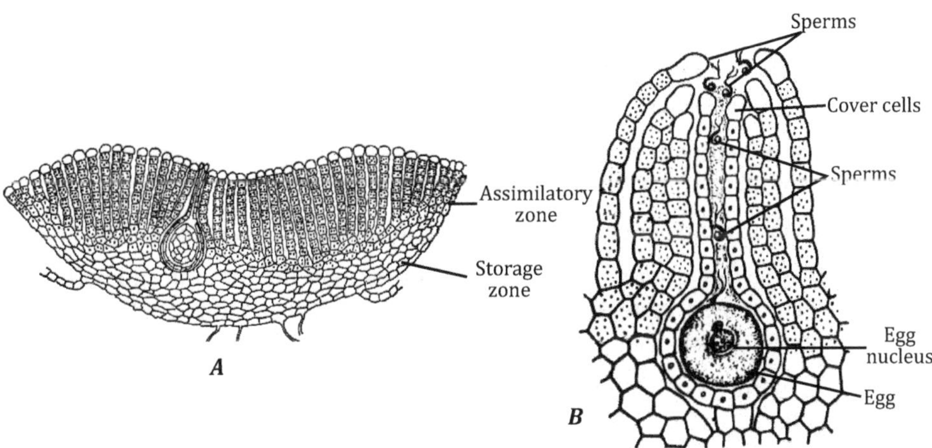

Figure 2.16*A&B Riccia* V.S. Thallus through archegonium. *A* - Shows a deep-seated archegonium. *B* - The archegonium has a long neck and a swollen venter. The antherozoids travel through the neck canal and reach the egg placed inside the venter bringing about fertilization.

forming a moist passage for the antherozoids to swim inside the venter and contact the egg cell. Post-fertilization event results in formation of two layered calyptra from the venter of the archegonium.

Fertilization

The antherozoids are produced inside the antheridium and the egg cell inside the archegonium. Fertilization ensues due to chemotactic attraction between the two and a zygote is formed. Thus, begins the sporophytic phase. Further mitotic divisions in the zygote result in multicellular embryo formation. Two major embryonic tissues are differentiated, the outer is the amphithecium (one celled thick) and a multicellular endothecium which is inner potentially fertile region. The protective jacket, also one layered in thickness arises from the amphithecium and endothecium gives rise to sporogenous mass which upon meiosis leads to haploid generation. The spores are set free upon disintegration of the capsule. In *R. crystallina,* some sporogenous cells form nurse cells, thus begins sterility in the sporophyte.

Sporophyte

- Sporophyte represented by a structure generally also referred to as sporogonium, is an undifferentiated mass of spores inside a unicellular jacket.
- Sporophytes of various age can be seen along the median groove with youngest placed towards the notch **(Plate II.5 and II.6)**.
- It is simplest among bryophytes; is spherical in structure and generally shows remnants of archegonial neck **(Fig. 2.17).**
- When young it acquires outer calyptra and inner calyptra that arise as a result of post-fertilization divisions of the venter wall **(Fig. 2.17A).**
- Sporogenous mass undergoes meiosis forming spore tetrads and later haploid spores **(Plate II.7A&B).**
- There are no elaters and therefore, the whole sporogenous mass is fertile.
- In *Riccia crystallina,* a few spore mother cells do not undergo meiosis to form spores, instead form sterile cells which act as nurse cells providing nutrition to developing spores.
- At maturity near spore tetrad stage, the single layered jacket disintegrates **(Fig. 2.17B).**
- The inner calyptra layer follows disintegration in much the same way as the jacket layer.
- Various stages of sporophyte development are seen along midrib **(Fig. 2.18A)**.
- The sporophyte is simple in construction. It is not differentiated into capsule, seta and foot. The spores are produced inside a rounded sporophyte which initially has a single layered jacket and two layered

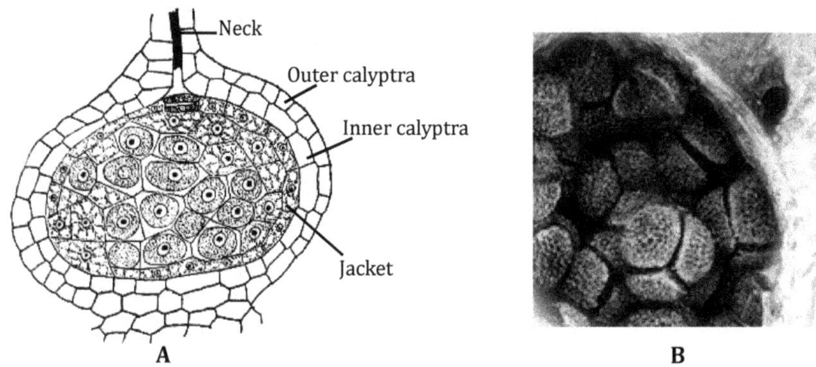

Figure 2.17*A&B Riccia* V.S. Thallus through a developing sporophyte. *A* - The spore mother cells are enclosed by jacket and two layered calyptra. *B* - Scanned micrograph of spore tetrad.

calyptra – inner and outer. Towards maturity the jacket and inner calyptra layer get disintegrated. The spores remain enclosed by outer calyptra. It must be noted that spores and calyptra are haploid. There is no diploid tissue in the sporophyte.

- The mature spores which are haploid (n) structures remain inside the sporophyte with only an outer calyptra (n) encasing them **(Fig. 2.18B).**
- Sporophyte lacks any special dehiscence or spore dispersal mechanism.
- The spore dispersal is by simple disintegration of the sporogonium (sporophyte).

Spores represent the first cell of gametophytic generation. Generally the spores are arranged in a tetrahedral manner. The wall is three layered, outermost

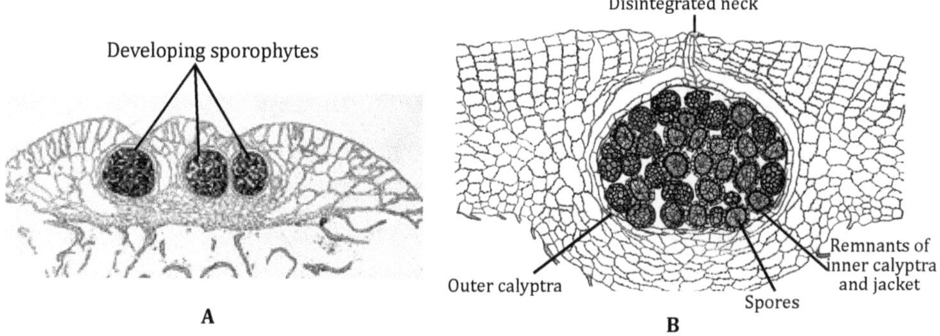

Figure 2.18*A&B Riccia* V.S. Thallus through sporophytes. *A* - To show the arrangement of sporophytes along the midrib. *B* - Spores enclosed by outer calyptra. Also seen are the remnants of archegonial neck and jacket.

is exosporium, which ruptures at the time of spore germination and a germ tube emerges. This tube elongates and then cuts off a few cells to organize an apical cell. Further growth gives rise to a flat dorsiventral gametophyte.

The salient features of the life cycle are depicted in **Figure 2.19.**

Evolutionary Trends (as proposed by Cavers 1910-11)

Riccia is an example of a bryophyte with both a simple gametophyte and a simple sporophyte. The sporophytic phase is represented by simple undifferentiated structure (sporogonium) which is without any seta and foot **(Plate II.7).** When young , sporophyte has a single layered jacket (haploid and sterile) encasing the

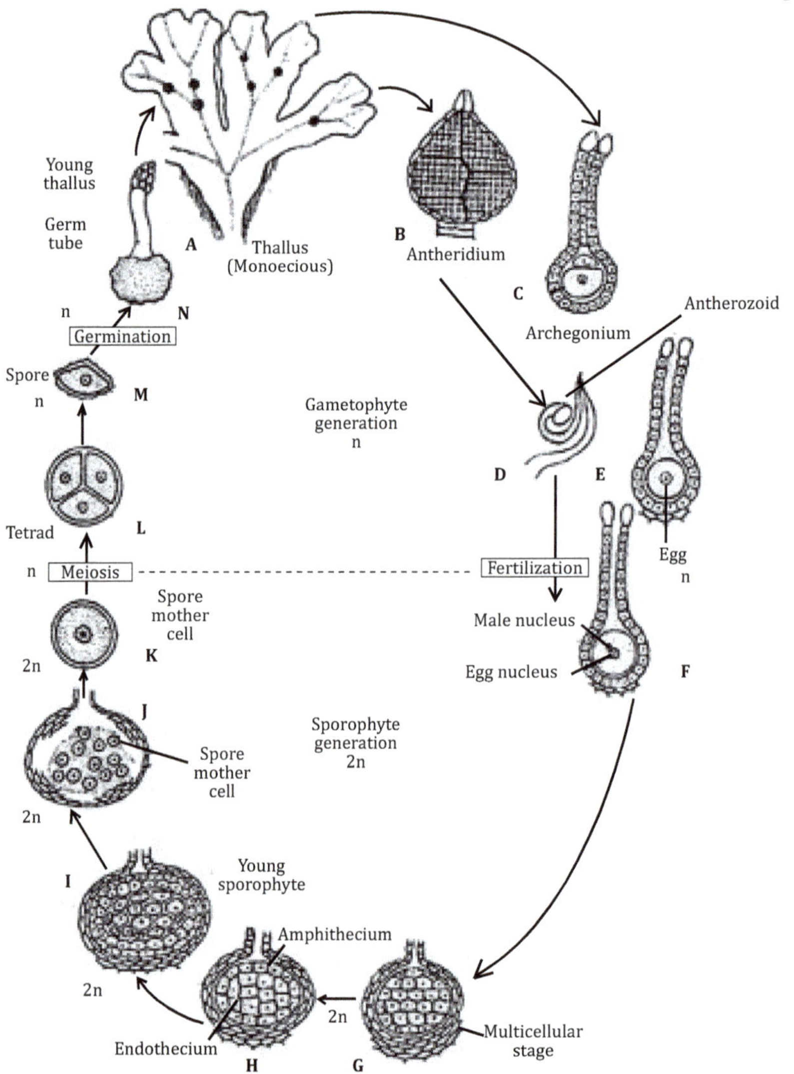

Figure 2.19 *Riccia* Life cycle of the liverwort.

mass of spores (haploid and fertile). Later, the sporophyte acquires two additional layers of protective coverings, the inner and outer calyptra (haploid and sterile). The sporophyte remains embedded in the gametophytic tissues. There is no elaborate mechanism of dehiscence and spores are liberated upon disintegration of the protective wall. In such a case, the spores fall in vicinity of each other and the parent plant, entering into competition for food, space and other resources. Many die due to the competition and only a few survive. The *Riccia* sporophyte was debated to be primitive in which sterilization had proceeded only as far as the formation of a single peripheral layer constituting the sporophyte wall; that the *Riccia* type of sporophyte was considered therefore, not only the simplest but the most primitive (Cavers 1910-11).

- In *Riccia,* the simplest sporophyte of any extant land plant is seen.
- There is no carpocephalum; the sporophyte (sporogonium) is embedded and virtually hidden in the gametophytic tissues. There is apparently no foot or seta (Schuster, 1992).
- At maturity, the spherical sporophyte consists merely of spores enclosed in outer calyptra as both the delicate unistratose jacket and inner calyptra are broken down. Any diploid structure is absent.
- Spores can be among the largest of any liverwort; these are typically thick walled, and long-lived.
- Spores are passively released upon decay of the capsule wall and surrounding thallus.

Riccia Vs *Marchantia* (As per previous classification, *Riccia* was placed before *Marchantia*)

Gametophyte

According to progressive evolution, the thallus of *Marchantia* is considered to be more complex. It shows following:

- The capillary system formed by the scales and rhizoids makes it more adapted for dry spells.
- The assimilatory zone with well-developed complex air pores and branched photosynthetic filaments add to the assimilative efficiency.
- Asexual reproduction through gemmae is an additional method of vegetative propagation.
- The aggregation of sex organs on raised stalks is a complex feature.

Sporophyte

- There is increase in the sterility of the sporophyte in *Marchantia*. The sporophyte of *Riccia* is simplest with no differentiation.

- In fact, the sporophyte of *Riccia* lacks any diploid structure and has spores (haploid) of new gametophytic generation and outer calyptra (haploid) of previous gametophytic generation. It is therefore, many a times debated that sporophyte is an erroneous term for *Riccia* **(Fig. 2.20).**
- Sporophyte in *Marchantia* has increased sterility where foot and seta are sterile structures. Further, inside the capsule, the elaters and apical cap are sterile. For one elater in *M. polymorpha* 128 spores are sacrificed.
- But this sterility has resulted in a more advanced and efficient system of capsule dehiscence and spore dispersal. The spores of *Riccia* are not dispersed far and wide, there is more competition for resources and many may not survive. However, in *Marchantia*, the twisting movements of elaters help spores be dispersed far off, thus, reducing the competition.

This lends greater survival chances.

** Maximum likelihood tree for 98 accessions of complex thalloids and 13 outgroups indicated that Riccia and Cyathodium have faster evolving rates and Marchantia with slower rate is more basally placed (Villarreal et al., 2015)*

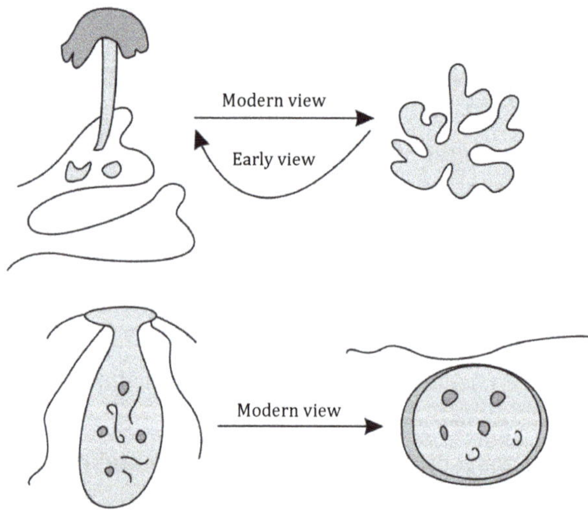

Figure 2.20: Evolutionary trend in Marchantiidae leads to reduction and simplification.

Marchantia (ancestral)	*Riccia* (more evolved)
• Compound air pores	• Simple air pore
• Single layered air chambers with basement filaments	• No air chambers but photosynthetic filaments present
• Ventral scales in more than two rows	• Ventral scales in one row
• Two or more rhizoid furrows in carpocephalum stalk	• No carpocephalum
• Spore: elater ratio greater than 4:1	• Elaters absent

** For recent views refer: Villarreal et al., 2016. Divergence times and the evolution of morphological complexity in an early land plant lineage (Marchantiopsida) with a slow molecular rate. New Phytologist: **209: 1734-1746.***

Do You Know?

- *Riccia fluitans* is an aquatic species which lacks rhizoids. It is sometimes called slender Riccia or Crystalwort, forms branching green ribbons about 0.1 centimetre (about 0.04 inch) wide and about 1.3 to 5 centimetres long that float in shallow ponds.
- Due to its versatility, it is often used for intricate aquascaping in show aquariums.
- *R. crystallina* shows nurse cells along with spores in the capsule.
- In *R. synspora*, *R. perssonii* and *R. curtisii* spores remain attached to each other even after tetrad stage and are dispersed in a group of four leading to formation of a group of four plants upon germination.
- **Riccia** is a commune (municipality) in the Province of Campobassoin the Italian region Molise, located about 15 km southeast of Campobasso, with a population of about 5,600.

Check Your Knowledge

1. Why is it erroneous to call *Riccia* sporophyte, a sporophyte?
2. *Riccia* sporophyte stands at the base of evolutionary ladder according to progressive evolution proposed by early works of Cavers. Comment.
3. According to theory of progressive evolution, *Riccia* is an example of a bryophyte nearest to the hypothetical ancestor and more complex sporophytes arose from it. Justify.
4. Explain the internal structure of thallus with the help of a neat diagram.
5. Give examples of dioecious and monoecious species of *Riccia*.
6. Which are the characters that make *Riccia* more advanced than *Marchantia?*

Class: Jungermanniopsida

- The gametophyte may be thalloid (*Pellia*) or differentiated into stem and lateral leaves (*Frullania*).
- In most cases, the gametophytes are without internal differentiation of tissues but certain genera have a central strand of thick-walled cells.
- Oil bodies where present, are one or more per cell. They are found both in gametophytic and sporophytic tissues.

- The ventral surface of gametophyte bears smooth walled rhizoids, scales being absent.
- The sex organs are generally found to be scattered on dorsal surface of thalloid members.
- The sex organs in leafy forms are produced on any branch of the gametophyte or on special branches as in leafy members.
- The archegonia arise from the young segments cut off by the apical cell and the apical cell may or may be not consumed in their formation.
- Neck of archegonium is usually formed of five vertical rows and there is no distinction between the neck and the venter, neck being as broad as the venter.
- The mature sporophytes lie some distance away from the growing apex of a gametophyte.
- Sporophyte is differentiated into foot, seta and capsule with a multistratose jacket.
- The capsule dehisces by four valves.

It has three subclasses: Jungermannidae, Metzgeriidae and Pelliidae.

Subclass Pelliidae

- Most genera that belong to this group, have a simple thalloid organization, a few may be leafy, and air chambers are altogether, absent. The archegonia are anacrogynous although exceptions occur.
- The subclass is morphologically heterogeneous. For example, all four types of apical cell geometries are expressed, with cuneate and lenticular types being of equal occurrence in the derived lineages (Shaw & Renzaglia, 2004).
- Organization of sex organs varies from being widely scattered, naked gametangia to tightly clustered within perigonia and perichaetia. Both male and female structures are borne on the dorsal surface.
- Sporophytes may be both large, massive and small, reduced types.

Order: Pelliales

- Plants are thalloid e.g., *Pellia* or leafy *e.g., Fossombronia* with succubous arrangement in the latter case.
- Apical cell is tetrahedral, cuneate, or hemidiscoid.
- Rhizoids hyaline or brownish to pale reddish brown, simple type.
- Antheridia arranged in two rows, or scattered or weakly clustered on the thallus, each in a conical or flask-shaped chamber with an apical ostiole or opening.

- Archegonia are arranged in two rows along the midrib as in *Noteroclada*, or in an acrogynous cluster, protected by a perichaetial flap or sheath (involucre) as in *Pellia*.
    - Sporophytes enclosed by a calyptra and caulocalyx as in *Noteroclada* or perichaetial pseudoperianth as in *Pellia.*
    - Capsules spheroidal, with conspicuous basal elaterophore, dehiscing by four valves; spore germination precocious and endosporic.

Family: Pelliaceae

- Plants are thalloid and the thallus is prostrate, dorsiventral and very often shallowly lobed by irregular incisions.
- Archegonia protected by involucre of gametophytic origin.
- Antheridia and archegonia remain scattered on the dorsal surface of the thallus.
- The capsule is globose or oval in shape. It possesses a basal elaterophore.

Pellia

Habitat and Distribution

Pellia Raddi., has robust thallus with broad, elongated lobes. It is worldwide in distribution and grows mainly in moist, shady places especially along ditches and streams. Sterile plants are seen growing under frequently running shallow water. Out of the six species reported world over (*P. neesiana, P. epiphylla, P. columbiana, P. megaspora, P. borealis* and *P. endiviifolia*), three (*P. neesiana, P. epiphylla* and *P. endiviifolia*) have been reported from India (Bapna & Kachroo, 2000) where two of these species (*P. endiviifolia* and *P. epiphylla*) have been recorded from Jammu and Kashmir state (Tanwir *et al.*, 2008). While *P. endiviifolia* is dioecious, *P. epiphylla* is monoecious.

Gametophyte

The spore germinates to form a filamentous structure that gives rise to a thalloid gametophore. *Pellia epiphylla* has medium-sized (often 1 cm or more wide), rather simple and sparingly branched thallus. The midrib is rather ill-defined, the sinous margin is not differentiated into a distinct wing, and the surface lacks an angular network of cells visible on the thallus as seen in *Marchantia*. Simple rhizoids arise from the under surface and are aggregated near the midrib; tuberculate rhizoids being absent. The thallus is internally simple, lacks intercellular spaces but is thickened near midrib (Asakawa, 2004).

Morphology

- Thallus is flat, dorsiventral, inconspicuously dichotomously branched and prostrate with sinuous margins **(Plate III.1).**

- The wings are thin and the midrib indistinct; the apical notch, however is distinct **(Fig. 2.21A)**.
- The ventral surface bears simple rhizoids more aggregated along the midrib.

Anatomy

- Internally thallus, unlike that of *Riccia* and *Marchantia* is simple, undifferentiated into photosynthetic and storage zones. It is compact and parenchymatous with upper and lower epidermis **(Plate III.2)**.
- The wings are thin while the midrib region is broad and multicellular **(Fig. 2.21B)**.
- The cells contain oil and sacculatane diterpenoids have been detected in *Pellia,* along with other genera - *Pallavicinia, Fossombronia, Trichocoleopsis,* and *Porella* (Asakawa, 2004).
- Simple rhizoids arise from the lower epidermis.

Reproduction

Vegetative reproduction is through death and decay or through gemmae and tuber formation.

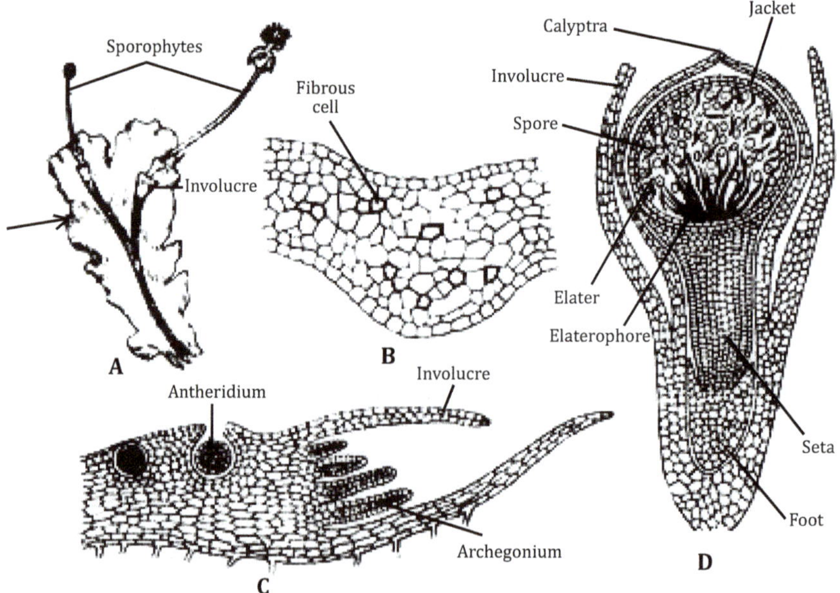

Figure 2.21 *A-D Pellia* Various stages in development. *A* -The sinuous margins of the thallus help prevent desiccation. *B* - V.S. Thallus shows internally a simple organization, with no distinction in photosynthetic and storage zone. *C* - Distribution of sex organs, formation of archegonia towards the tip and antheridia towards the dorsal surface. *D* - The mature sporophyte with foot, seta and capsule which shows a basal elaterophore.

Sexual Reproduction

Pellia epiphylla is monoecious, with male organs scattered in a wart-like broad row along the midrib. A vertical flap or involucre forms a pocket-like structure in which the female organs are formed **(Fig. 2.21A).** The archegonia are formed in groups, where young and old archegonia are indiscriminately intermingled **(Plate III.A).**

Antheridia

- In the monoecious or bisexual species as in *P. epiphylla*, the antheridia are borne in cavities on the dorsal surface of the thallus along the midrib **(Fig. 2.21C)**.
- A single antheridium resides in the antheridial cavity which opens to the outside by a small pore.
- The nearly spherical antheridium is attached to the base of the cavity with a small multicellular stalk. A single layered jacket encloses numerous antherozoids at maturity.

Archegonia

- The group of four to twelve archegonia arises just behind the apical notch.
- There is no particular arrangement as both old and young archegonia remain mixed. The entire group is covered by a flap-like involucre that arises from the thallus tissue.
- The archegonia with a less dilated venter and a neck remain attached to the thallus with a small multicellular stalk.
- The neck is formed of five vertical rows of cells and usually six to eight neck canal cells are present.

Fertilization

Takes place when the thallus is wet. The male sex organs (antheridia) absorb moisture and release the antherozoids which are chemotactically attracted to the archegonium. Presence of moisture helps them swim through the canal formed by dissolution of neck canal cells and venter canal cells.

Sporophyte

The mature sporophyte is a well-defined and a complex structure consisting of foot, seta and a capsule **(Fig. 2.21D).**

Foot

- Is a conical and parenchymatous structure that remains anchored in the gametophyte tissues **(Plate III.4).**
- The distal end of the foot extends like a collar around the base of the seta.

Seta

- It is composed of non-chlorophyllous parenchymatous cells arranged in definite longitudinal rows.

- In a young sporophyte, the cells of the seta are relatively small and contain abundant starch, but at maturity the cells elongate rapidly and become are devoid of starch.

Capsule

- It is almost spherical structure with a two or more layered jacket or capsule wall.
- The cells of the outer layer of the jacket are polygonal with radial thickening bands.
- The cells of the inner layer are relatively thin and have semilunar thickening bands, which help in the dehiscence of the capsule.
- The jacket encloses a mass of spores and elaters. Elaters have a double spiral band of wall thickenings on the inside of the cell wall. They are long slender and spindle shaped.
- A characteristic feature of the sporophyte is the presence of a basal elaterophore.
- Elaterophore consists of 20-100 elongated elaters which radiate from the base of the capsule cavity towards the apex **(Fig. 2.21D, Plate III.4B).**
- Each spore is unicellular and haploid, formed by lobing of spore mother cells.
- The young sporophyte is covered by a protective sheath, the calyptra which develops from the cells of the venter wall.
- The sporophyte develops inside the involucre - a membranous structure formed from the gametophytic tissue. The sporophyte at maturity ruptures the involucre which after capsule dehiscence stays behind like a pouch.
- As the mature capsule dries up, the wall splits into four valves along the four lines of dehiscence. The seta elongates and the four valves of the dehisced wall become reflexed. The elaterophore is now exposed and the elaters exhibit hygroscopic movements, dispersing the spores **(Plate VIII.7A-C)**.
- Spores show precocious germination and the sporelings at the time of liberation may either be 4-celled or 6-celled. The multicellular sporelings do not undergo any period of rest and develop immediately after falling on substratum. They develop an apical cell and a thalloid plant is formed by the activity of the apical cell.

The life cycle is depicted in **Figure 2.22.**

Evolutionary Trends (as proposed by early works)

Though the gametophyte has remained very simple externally as well as internally in *Pellia,* but the sporophyte continues to show progressive sterilization in

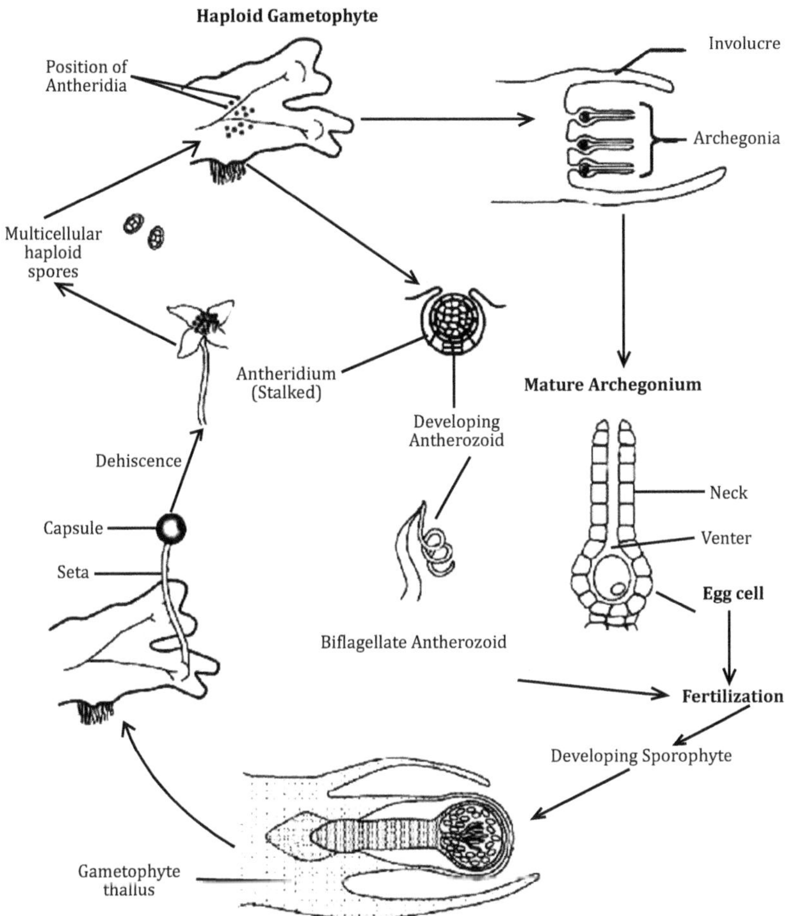

Figure 2.22 Life Cycle of *Pellia*.

tissues. From sporophyte of *Riccia* where the sterile tissues are jacket and nurse cells (in *R.crystallina*), the spore bearing structure in *Pellia* shows greater sterility in the form of foot, seta, elaters, the multistratose jacket and the elaterophore. However, the dispersal mechanism is more efficient and so even if the spores produced are less in number when compared to *Riccia*, they are dispersed more efficiently.

**The representatives of the genus Pellia, are recognized as the most basal lineage of the simple thalloid liverworts with respect to many plesiomorphic features, such as cuneate apical cell, a thallus without a midrib, a spherical capsule and massive seta (He-Nygrén et al. 2006; Crandall-Stotler et al, 2009). Identification of miRNAs in Pellia and their high similarity to algal miRNA and sRNA sequences, confirms the location of P. endiviifolia at the root of the land plant evolutionary tree of life.*

Do You Know?

Pelleas or **Pellias**, Sir Pelleas was a Knight of the Round Table in Arthurian legend and was one of the six knights who could defeat Gawaine.

Markren Pellias was the Planetary Trade Director (The **Planetary Trade Director** was an elected official whose job was to ensure fair trade for people on the planet Rhamalai).

(Rhamalai was discovered and settled by the quasi-religious sect of Cherishites, approximately four hundred years before the Galactic Civil War (http://starwars.wikia.com/wiki/Planetary_Trade_Director)

Round *Pellia* (*Lomariopsis lineata*) is an awesome looking fern and great for shrimp to feed from or fish fry to hide in. This plant was first misidentified as liverwort *Pellia endiviifolia* not before the molecular phylogenetic analysis determined it to be a fern gametophyte. It is now known as **Süsswassertang** and is a type of aquarium plant (https://en.wikipedia.org/wiki/Süsswassertang).

Check Your Knowledge

1. Give a detailed account of the gametophyte of *Pellia*.
2. Discuss the structure of sporophyte of *Pellia*.
3. Define: elaterophore, involucre.

Subclass Jungermanniidae

- Commonly referred to as "Leafy Liverworts" with leaves being isophyllous or anisophyllous.
- The group is diverse, composed of nearly 4000 to 6000 species distributed in temperate, subtropical, and tropical regions.
- The plants are typically well-branched and form mats across the substratum.
- Three rows of leaves are seen, with two rows of lateral leaves which are similar and lobed, and a row of ventral leaves also known as amphigastria, which are highly reduced or may be absent all together in some forms.
- Sex organs are present on branches with antheridia in axils of modified leaves and archegonia usually terminal and remain surrounded by perianth formed by coalescing leaves.
- Capsule wall is multistratose.

Order Porellales

- All members of this order are prostrate, dorsiventral and leafy.
- The leaves may be in three rows where the third row is of small leaves or

the leaves are highly reduced as in *Frullania*. In *Radula* sp the third row of leaves is altogether absent and only two lateral rows of leaves are seen. These leaves are closely placed and overlap frequently.

- Leaves show incubous arrangement - where the leaves are so arranged that the upper margin of each leaf lies above the lower margin of the next leaf *e.g., Frullania, Porella* and *Radula*.
- The antheridia are borne in axil of specialized leaves either singly or in groups.
- The archegonia are always apical in position and the apical cell is consumed in their formation. As a result, even the sporophyte which dehisces along four lines, is terminal in position.
- Elaters co-exist with spores that arise from lobed spore mother cells. Two families Porellaceae and Frullaniaceae are discussed with detailed life cycle of one representative each.

Suborder: Porellineae

Family Porellaceae

- Leaves two lobed, are incubous and the postical lobe is distinct; amphigastria present.
- Rhizoids form tufts at the base of underleaves.
- Sex organs arise on short branches.
- Capsule is spheroidal, splits only half way down.

Porella

Habitat and Distribution

The genus has a wide distribution in temperate areas, where it is commonly found growing attached to the bark of trees. Besides, it colonizes deep and humid valleys and slopes and streams, as well as exposed rocks at the highest peaks. *Porella platyphylla* is a typical member of the genus. *Porella pinnata, P. bolanderi, P. vernicosa* and *P. rivularis* are some of the common species.

Gametophyte

The spore germinates to produce a small protonema-like structure from which an apical cell gets established. The cell divides and re-divides to form a dorsiventral leafy axis which is much branched, branching being monopodial. The stem bears two rows of lateral leaves arranged in incubous manner. These leaves have a larger antical lobe and a smaller postical lobe. Amphigastria are small scale-like leaves on the ventral surface. Simple rhizoids on the ventral surface help in water conduction and anchorage.

Morphology and Anatomy

- The gametophyte is dorsiventral, foliose and grows flat on the substratum.
- The axis or the stem is bi- or tripinnately branched. It bears leaves in three rows, two lateral and one is ventral **(Fig. 2.23A).**
- The lateral leaf is bilobed, the upper anterior lobe called antical lobe, is larger.
- The lower lobe known as postical lobe or lobule is much smaller and has an acute apex.
- The lateral leaves shows incubous arrangement *i.e.,* lower edge of each leaf is covered by the upper edge of leaf below **(Fig. 2.23B).**
- The leaves of the ventral row called amphigastria are small and resemble the postical lobe of the dorsal leaves, but differ from them in being broader with decurrent bases.
- The margins of the leaves may be entire.
- A large number of unicellular smooth walled rhizoids arise from the ventral surface of the stem which lacks any conducting tissue but has only parenchymatous cells with little differentiation into cortex and medulla.

Reproduction

Vegetative reproduction occurs by gemmae formed generally on the lower surface of leaves. The leafy axis shows great regenerative capacity.

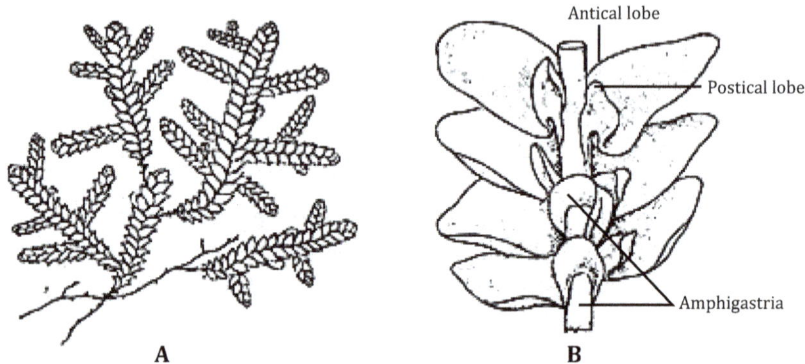

Figure 2.23*A&B Porella. A* - Thallus to show three rows of leaves. *B* - The two lateral rows of leaves show incubous arrangement while the ventral leaves are modified into amphigastria.

Sexual Reproduction

Separate male and female plants are present in dioecious species of *Porella*. The sex organs are formed on special branches known as antheridial and archegonial branch **(Plate III.5&6).**

Antheridial Branch

- The antheridial branches project at about right angles from the main branch.
- The leaves known as bracts or perigonial leaves are closely placed and each bract bears a single antheridium in the axil **(Plate III.6).**
- The antheridium has a long two celled stalk and a jacket which is two layered at the base and single layered distally **(Plate III.7).**
- Biflagellate antherozoids are produced inside the antheridium.
- At maturity, the antheridial cells absorb water and the thinner upper part bursts open into number of irregular lobes which curl back releasing the antherozoids *en masse.*

Archegonial Branch

- Archegonia are present terminally on specialized archegonial branches of the female plant **(Plate III.8).**
- Apical cell of the female branch divides producing initially a few segments that develop into leaves or bracts while the later derivatives as well as the apical cell function as archegonial initial.
- The apical cell of archegonial branch is eventually used up in formation of last archegonuim and further growth of archegonial branch ceases.
- The archegonial branches which are shorter than the vegetative branches have only two to five leaves or bracts or the perichaetial leaves. The lower bracts form an involucre, whereas two uppermost bracts coalesce to form perianth.
- The latter encloses a group of ten to fifteen archegonia.
- Archegonia develop in acropetal succession where the mature archegonium is at the base and youngest at the tip of archegonial head.
- The nearly mature archegonium has a neck which is almost as broad as venter. The neck consists of five vertical rows of cells and inside the neck there are six to eight cells (neck canal cell). Venter has a two-layered wall and encloses a small egg and ventral canal cell (VCC).

Fertilization

Occurs as in other liverworts being dependent on water. Fertilization leads to zygote formation which gives rise to an embryo, and later the sporophyte develops from the embryo.

Sporophyte

- The zygote secretes a wall and begins to increase in size. It then divides by transverse walls forming a three celled structure. The lowermost cell does not contribute to the mature sporophyte and instead forms a suspensor.

The whole sporophyte with spore bearing capsule, seta and indistinct foot is derived from the other two cells.

- The capsule is globose with multistratose jacket or wall **(Fig. 2.24).**
- The cells of the sporogenous region are arranged in more or less marked rows radiating from the base of the capsule. Some of the sporogenous cells stop dividing and grow regularly in all directions differentiating into spore mother cells, While the other sporogenous cells may divide further and elongate to form the sterile elaters.
- The spore mother cells (2n) become four-lobed before they begin to divide; the nucleus divides by reduction division to form four haploid spores (n).
- The young sporophyte is borne at the tip of the female branch and is surrounded by three definite envelopes - the calyptra, the perianth and the involucre **(Plate III.9)**.
- Calyptra is several-layered thick. It encloses the capsule from the outside and ruptures at maturity to facilitate release of spores.
- Involucre is formed by the enlarged bracts surrounding the base of perianth.

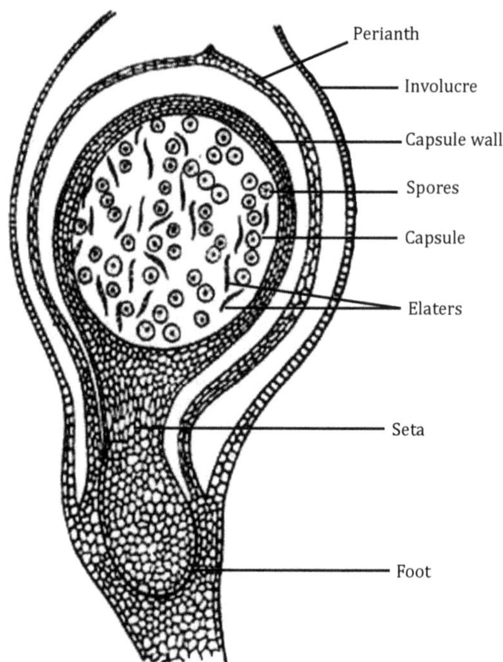

Figure 2.24 *Porella*. L.S. Sporophyte with foot, seta and capsule. The capsule wall is multilayered and the spores are intermixed with elaters. Also seen are perianth and involucre, both of gametophytic origin.

Seta

- Seta is short and merges gradually into the base of the capsule and is made up of two or more than two layers of cells.
    - It is a link between the capsule and foot of the sporophyte. When the spores are mature, seta elongates pushing the capsule out from its protective envelopes, thus helping in dehiscence of the capsule.

Foot

- The foot remains less conspicuous and is simply somewhat enlarged portion of the seta hence the distinction between these two is not very clear.
- The foot grows downward into the tissue of the stem of the archegonial chamber or branch.
- The foot remains embedded in a short pouch formed from the outer tissue of the stem of archegonial branch **(Plate III.9).**

The salient features of the life cycle are depicted in **Figure 2.25.**

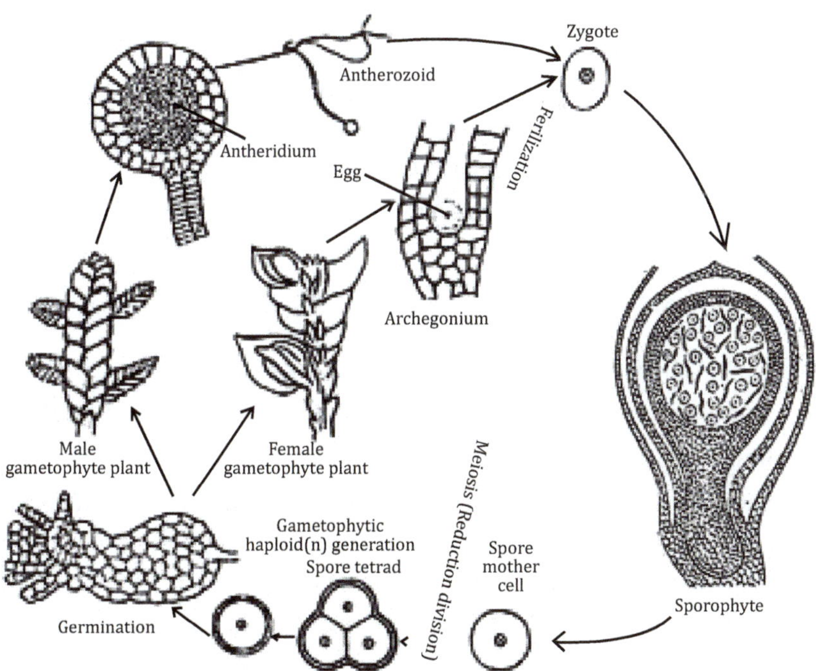

Figure 2.25 ***Porella*** **Life cycle of the foliose liverwort.**

Evolutionary Trends (as given in theory of progressive sterilization Cavers 1911).

The leafy liverwort has an efficient water retaining system provided by overlapping leaves and the rhizoids.

There is increased sterility in the sporophyte. To begin with a portion of the embryo is given to suspensor formation, and the entire sporophyte is obtained from the $2/3^{rd}$ of the embryo. In the fertile region of the capsule, the sterility increases with multistratose nature of the wall and presence of elaters, the formation of which uses up a portion of the sporogenous tissue.

Do You Know?

Paella, a dish made from rice and vegetables originated in Spain and must have been derived from the phrase *por ella* meaning for her (According to a legend a lover once prepared it for his fiancée). In the second meaning it is a pan with two handles used to prepare this dish (https://books.google.co.in/books?id=CDwpCgAAQBAJ and pg=PT68 and lpg=PT68 and dq).

Vivo per lei (English: "I Live for Her") is the name of a 1995 song recorded by an Italian artist Andrea Bocelli as duet with Giorgia for his album *Bocelli.* https://en.wikipedia.org/wiki/Vivo_per_lei-.

Check Your Knowledge

1. Explain water absorption and conduction amongst Marchantiophyta.
2. Tabulate the similarities and differences of gametophyte in the four genera of Marchantiophyta studied by you.
3. Taking examples from liverworts, discuss the complexity trends in:
   a. Gametophyte
   b. Sporophyte

Suborder: Jubulineae

- Plants usually with underleaves (absent in a few Lejeuneaceae).
- Rhizoids arranged in fascicles from the underleaf base.
- Sporophytes enclosed by a stalked, true calyptra and beaked perianth.
- Capsules are spheroidal, with two layered wall; elaters are vertically aligned and attached to the valve apices.
- Spores possess rosette markings in the exine (absent in *Jubula*).

Frullaniaceae

- Stem is pinnately branched and branch develops from the ventral half of the lateral segment replacing the lobule of the leaf. This is the Frullania type (Bryopteris-type on decapitated shoots) branching with incomplete leaf in between the branch and the higher axis.

- Leaves are 3-lobed, with the dorsal or antical lobe usually entire, the postical lobe or the lobule forms a water sac, a laminar stylus is seen at the base of postical lobe; amphigastria are bifid.
- Antheridia are formed on elongate to capitate Frullania-type branches, occasionally becoming intercalary.
- Archegonia arise on leading axes, with multiple archegonia; perichaetial leaves in 3 or 4 series. Perianth formed from uppermost perichaetial leaves has a compressed mouth. In some cases the arrangement is interpreted as - perianth, is usually surrounded at its base by modified leaves, called **bracts**, and modified amphigastria, called bracteoles.
- Seta up to twelve cells in diameter, non-articulate.
- Gemmae are absent.

Frullania

Habitat and Distribution

The liverwort genus *Frullania* Raddi, with a worldwide distribution, is considered large and complex comprising over 1000 described species (von Konrat & Braggins, 2001). *Frullania tamarisci* can be found on dry granite cliffs or occasionally on bark of deciduous and coniferous trees and on rotting logs.

Gametophyte

The spore germinates to form protonema-like structure which after a few weeks gives rise to a fragile and branched stem. The plants are reddish brown or purplish, seldom green, not very small, dorsiventral and leafy.

Morphology

- The main plant body is a dorsiventral leafy axis. The prostrate stems are pinnately branched.
- On the stem are three rows of leaves - two rows of lateral leaves and a row of underleaves or amphigastria.
- The lateral leaves are arranged incubously. Each leaf has an antical lobe and a postical lobule. The antical lobe is entire and ovate while the postical lobule is modified into a pitcher.
- These pitchers are bladders in which water enters through capillarity. All species may not form pitchers and the lobule remains open. Where the pitchers are formed, their role is to collect rain water. Even the space between stylus and stem holds water while the antical lobe continues to photosynthesize.

- The postical lobule has a short stylus near its attachment.
- The underleaves or amphigastria are rounded or notched **(Fig. 2.26)**.

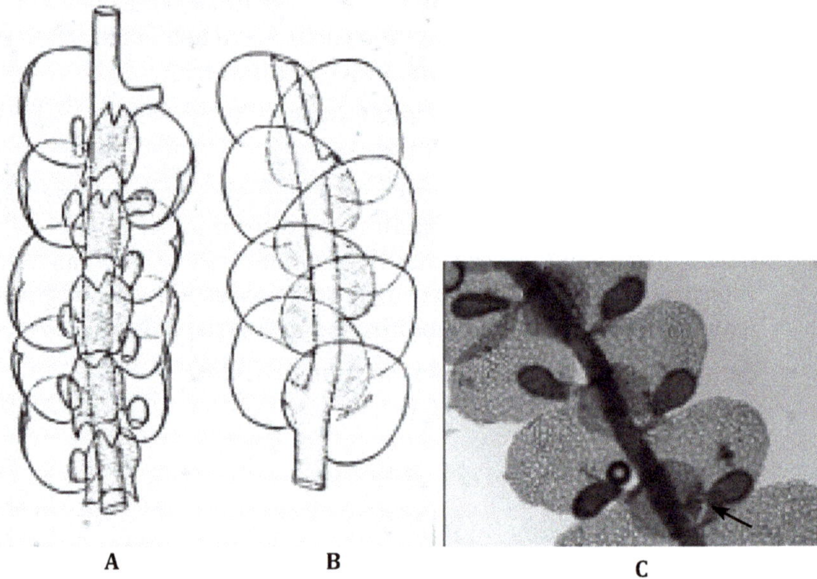

Figure 2.26*A-C Frullania. A*- Plant shows three rows of leaves. *B* - The two lateral leaves are incubous. *C* - They are complicatedly bilobed, divided into a large dorsal lobe, a smaller ventral lobule, and a small stylus (arrow) between lobule and stem. Under leaves are equally bilobed.

Anatomy

- Stem shows outer cortical zone and inner medullary zone.
- Leaf has a single layer of cells with each cell bearing several chloroplasts and oil bodies.

Reproduction

Vegetative reproduction occurs through gemmae.

Sexual Reproduction

Plants may be dioecious or monoecious. Sex organs are borne on respective branches.

Antheridia

- Antheridia are borne in axil of bracts, the specialized perigonial leaves. The bracts have two equal lobes. There are two antheridia in axil of a bract.
- Mature antheridium has a slender stalk and a globose body.

Archegonia

- Two or three archegonia are formed at the apex of the archegonial branch and are surrounded by two to five pairs of perichaetial leaves, the uppermost bracts fuse to form perianth.
- The mature archegonium is flask-shaped and the neck is typical of the group.

Sporophyte

- It has a blunt foot for anchorage and absorption of water.
- The seta is short and gives way to capsule which has a multicellular wall where the cells show differential thickenings. A number of spores and elaters are present inside the capsule. The elaters are mostly attached to the inside of capsule wall.
- When spores reach maturity, seta elongates and pushes the capsule out of calyptra and perianth.
- The capsule dries up, the thin layers of the capsule wall also dry up but the thick walls do not. Due to the difference in the water status of these walls, a tension is created.
- Under this unequal pressure, the capsule dehisces along four lines of dehiscence from top down forming four valves with elaters attached to the tip on one end and to the base on the other.
- Due to the tension, the elaters are freed at the base and this makes them contract and swing flicking away the spores through their sling-like action.
- The spores fall on the substratum and germinate to produce a new gametophyte.

Chapter 3

Phylum – Bryophyta – *The Mosses*

Bryophyta commonly referred to as 'mosses' with approximately 13,000 species (Hyvönen *et al.*, 2004), is the third most diverse group of land plants (other two being ferns and angiosperms). Mosses share with Marchantiophyta and Anthocerotophyta, a haplodiplobiontic life cycle. The leafy gametophyte is free-living and autotrophic, while sporophyte develops into an unbranched axis with a terminal spore-bearing capsule. The sporophyte remains physically attached to the gametophyte and is at least partially physiologically dependent on the gametophyte. Recent phylogenetic reconstructions suggest that three lineages, (liverworts, mosses and hornworts) of early land plants compose an evolutionary grade that spans the transition to land and the origin of plants with branched sporophytes as seen in angiosperms. The mosses seem to occupy an intermediate position as their origin is early to the divergence of the ancestor to the hornworts and vascular plants but they have evolved from a common ancestor with liverworts.

This group of plants (mosses) includes the most advanced bryophytes. They are strongly adapted amongst bryophytes to terrestrial environments, are cosmopolitan and flourish well in moist and damp places. They have highly developed mechanisms of desiccation tolerance (Oliver *et al.,* 2005; Proctor *et al.,* 2007) and, unlike liverworts, show considerable diversity in their sporophytes. *Fontinalis antipyretica*, an aquatic moss grows deep in the lakes while *Tortula moralis* grows on exposed rocks in dry places. The moss gametophyte differs from that in liverworts because two distinct stages are recorded in gametophytic phase – the protonemal and the gametophore stage. In liverworts protonemal stage is short-lived indistinct and represented by a germ tube only and is frequently called sporeling. In mosses, the sporophyte is long-lived, complex in construction and the dehiscence of capsule and spore dispersal in highly evolved forms, is a specialized mechanism assisted by peristome teeth.

Salient Features

- It represents the largest division under bryophytes.
- Gametophyte is differentiated into two distinct stages: a branched prostrate, juvenile filamentous or thalloid stage, protonema which (compared to the other phase) is generally short-lived and an erect, leafy dominant and long-lived gametophore.
- The gametophore has a stem with spirally arranged leaves and remains attached to substratum by rhizoids.
- The rhizoids which resemble protonema excepting absence of chloroplasts, are multicellular, branched and possess oblique septa.
- During the reproductive stages, sex organs are formed from the superficial cell of the gametophore and frequently are interspersed with paraphyses.
- During development, sex organs organize an apical cell.
- The sporophyte is a well differentiated structure and shows determinate growth.
- The jacket or capsule wall is multistratose and functional (*Funaria*) or non-functional (*Sphagnum*) stomata span the epidermis.
- The archesporium, representing the fertile region, develops from the endothecium or amphithecium and the centre is occupied by columella that arises from endothecium.
- Dehiscence of capsule and spore dispersal is a complex process.
- A distinct structure, protonema is formed on spore germination. Protonema is a highly branched filamentous uniseriate structure. Cell specialization occurs to form two types of filaments in protonema - a horizontal system of reddish brown anchoring filaments called the caulonema and upright, green filaments, the chloronema.
- Each protonema can spread over several centimeters forming a fuzzy green film over the substratum. As protonema grows, certain cells usually on the caulonema generate bud initials. Mostly, numerous shoots develop from each protonema.
- This group of bryophytes is diverse with a range of complexity displayed **(Table 3.1)**.

Table 3.1 Characteristic features of some classes of Bryophyta

Class	Characters	Order
Takakiopsida	Capsule raised on a seta that is composed of sporophyte tissue. Capsule opens via a single, longitudinal, spiral slit.	Takakiales (1 family, 1 genus, 2 species).
Sphagnopsida	Capsule raised on a pseudopodium that is composed of gametophyte tissue. Capsule opens by means of an operculum.	Sphagnales (1 family, 1 genus, between 100 and 300 species). Ambuchananiales (1 family, 1 genus, 1 species).
Andreaeopsida	Plants are autoecious. Capsule raised on a pseudopodium. Capsule opens along several vertical dehiscence lines with tips of intervening segments remaining attached to one another at poles.	Andreaeales (1 family, 2 genera, under 100 species).
Andreaeobryopsida	Plants are dioecious. Capsule raised on a seta. Capsule opens along several vertical dehiscence lines. Valves are free, not jointed at tips.	Andreaeobryales (1 family, 1 genus, 1 species).
Polytrichopsida	Capsule raised on a seta. Capsule opens by means of an operculum. Capsule with peristome teeth (except for one genus). Peristome teeth nematodontous (See definition below). Costa/Nerve present in leaf (except for one genus).	Tetraphidales (3 families, 4 genera, no more than about 50 species). Polytrichales (1 family, 23 genera, several hundred species). Gametophyte typically very robust. Leaf with nerve and typically with lamellae along the nerve.
Bryopsida	Capsule sessile or raised on a seta. Capsule generally opens by means of an operculum, but by disintegration in a very few genera. In most genera the capsules have peristome teeth. Peristome teeth arthrodontous (See definition below). Gametophytes small to robust. Leaves with or without nerves.	16 orders, (Funariales in syllabus), 107 families, about 880 genera, about 12,000 species.

(Reference for classification in mosses is based on work of Buck, WR &Goffinet, B. (2000). Morphology and classification of mosses, pp 71-123 in Shaw, AJ & Goffinet, B. (eds) (2000). *Bryophyte Biology* Cambridge University Press).

* Also refer to: Systematics of the Bryophyta (mosses): from molecules to a revised classification by Goffinet, B and Buck R. W. 2004.

* Nematodontous peristome consists of whole dead cells while arthrodontous peristome teeth are differentially thickened or ornamented.

Occasionally, the mosses are categorized into eight classes **(Fig. 3.1)** - Takakiopsida (Takakiales, Takakiaceae), Sphagnopsida (Sphagnales, Sphagnaceae and Ambuchananiaceae), Andreaeopsida (Andreaeales, Andreaeaceae), Andreaeobryopsida (Andreaeobryales, Andreaeobryaceae), Oedipodiopsida (Oedipodiales, Oedipodiaceae), Polytrichopsida (Polytrichales, Polytrichaceae), Tetraphidopsida (Tetraphidales, Tetraphidaceae) and Bryopsida (21 orders and 104 families).

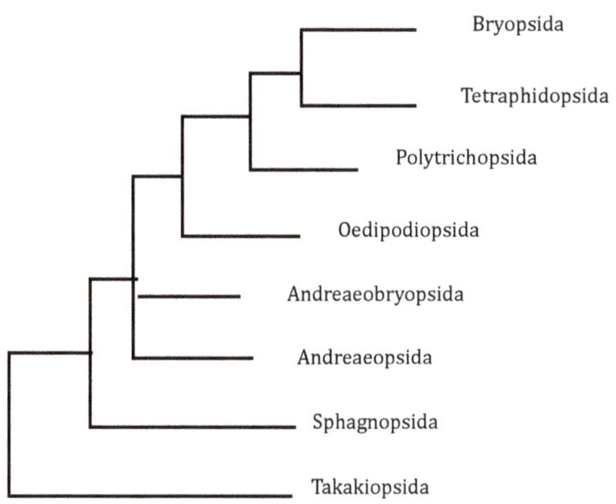

Figure 3.1 Phylogenetic relationships between basal groups of mosses based on molecular data.

Subphylum Sphagnophytina Doweld

Class Sphagnopsida Ochyra

The Sphagnopsida, (after Goffinet *et al.*, 2008) comprise a morphologically distinctive group of mosses.

Within the mosses, Sphagnopsida are distinguished from other mosses by numerous features.

- Gametophyte features include thalloid protonemata (shared with Andreaeopsida and a few early diverging members of the Bryopsida), presence of dimorphic leaf cells at maturity, and absence of rhizoids in mature plants.
- Enlarged, hyaline leaf cells, dead at maturity and with thickenings and pores, are enclosed within a network of much narrower chlorophyllose cells. This pattern is not found in any other moss, although loosely analogous cell dimorphism occurs in scattered species of the advanced mosses.

- Ontogeny of the chlorophyllose and hyaline cells during leaf development in *Sphagnum* is unique.
- Sporophyte generation of Sphagnopsida is also highly differentiated. The sporophyte lacks a seta (stalk) and is instead raised on a pseudopodium of (maternal) gametophyte origin.
- The sporogenous tissue within the capsule develops from embryonic amphithecial rather than endothecial tissue and overarches a massive columella (derived from the endothecium).
- The columella is much more slender in other mosses and the sporogenous tissue is derived from the outer layer(s) of endothecium, rather than the amphithecium as in other mosses.
- Endothecium and amphithecium are early to differentiate during sporophyte development.
- The capsules of Sphagnopsida dehisce by an apical operculum, as in Bryopsida (in contrast to the vertical or spiral suture lines as seen in Andreaeopsida/Andreaeobryopsida and Takakiopsida, respectively). However, unlike most true mosses Sphagnopsida have no peristomes and the capsules open via a unique "pop-gun" mechanism wherein the operculum detaches explosively as the capsule shrinks in diameter.

Order Sphagnales Limpr.

Family Sphagnaceae

- The group usually exhibits characters which may not be at times typical of mosses.
- Thalloid protonema is one celled thick and attached to the substratum by rhizoids typical of mosses.
- Only one gametophore arises from the protonema.
- The gametophore is leafy and leaves lack a midrib but possess two types of cells.
- The branches on the gametophore arise in fascicles.
- The antheridia are lateral and archegonia terminal in position.
- Amphithecium gives rise to archesporium while the entire endothecium gives rise to columella which never grows through the spore sac.
- Capsule opens by an operculum. There is no peristome.
- The capsule is raised by pseudopodium (comparable to seta in other mosses).

Sphagnum - The Peat Moss

Habitat and Distribution

Sphagnum (peat moss), with 200–300 species worldwide, comprises several large clades, classified as subgenera (Shaw *et al.*, 2010). *Sphagnum* moss dominates the plant cover of boreal bogs and accumulates carbon as peat. However, discoloured necrotic *Sphagnum* patches are also common in bogs. The northern hemisphere regions, *e.g.* Canada and North America, appear to be the centres of dominance. Spread of the moss extends from the tropics and north and south temperate zones to subarctic and antarctic regions (Parihar, 1965).The species are aquatic or semiaquatic and inhabit the areas with low pH. They grow in dense masses or cushions in swamps, ponds and lake margins, moist heaths and wet hill sides and may rarely grow in patches. In colonies, thousands of plants form a low blanket, covering large areas of land. The moss thrives in waterlogged, acidic areas where only a few other plants can survive.

Gametophyte

The plant begins its gametophytic phase as a haploid spore. The unicellular spore germinates soon after release to produce a germ tube leading to a filamentous protonema which transforms into a multicellular, but unistratose thalloid structure. This structure forms a bud opposite to the rhizoids and gives rise to a leafy gametophore which is unique in its organization. The plant is pale green with shades of pink, red, yellow or brown. Individual specimens are only a few inches tall and consist of a single central stem with short side branches bearing tiny, closely packed leaves, and young plants remain anchored by rhizoids typical of mosses.

The gametophore after a period of vegetative growth bears male and female branches on which are borne antheridia and archegonia respectively. Fertilization leads to zygote formation and the diploid sporophytic phase begins. The embryo after organizing two fundamental tissues, endothecium and amphithecium forms a sporophyte with foot, indistinct neck represented by a constriction and a spherical capsule. The sporophyte remains on a stalk-like structure - the pseudopodium and dehisces explosively, dispersing spores far and wide.

During dessication, the growth of bog moss is checked and the physiological activities like respiration and photosynthesis are suspended. The cytoplasm shows a high degree of resistance to dessication. When water is available physiological activities are resumed and normal growth of the plant sets in. Such plants are known as pallacuophytes (Buch, 1947).

Morphology

- Plant is erect perennial gametophyte with stem and leaves **(Fig. 3.2).** The stem being weak individually, gains support by aggregation and grows in dense masses or cushions. In reproductive phase, cluster of sporophyte is seen at tip of the axis **(Plate IV.1)**.

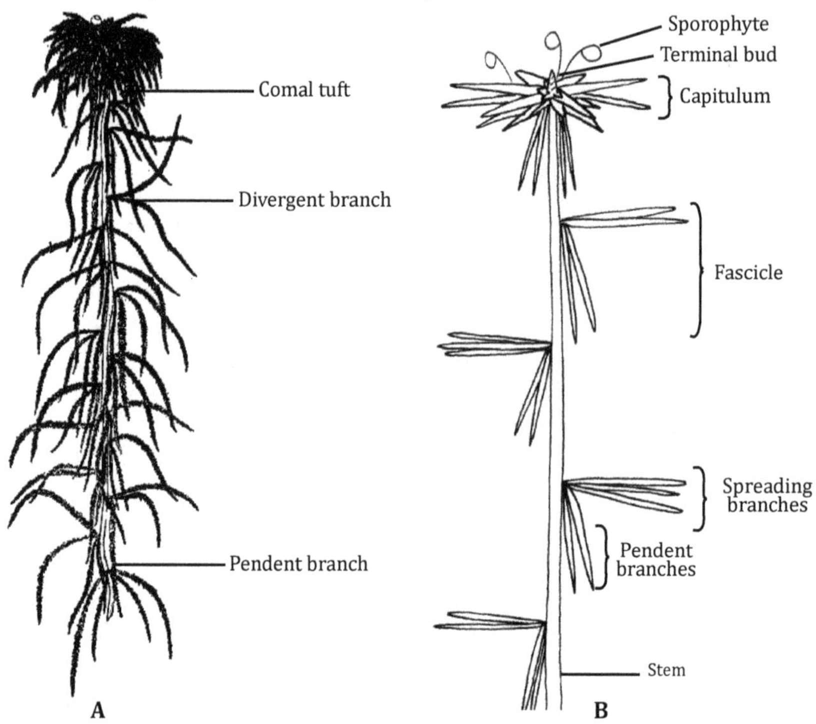

Figure 3.2 *A&B Sphagnum. A* -The plant shows a leafy axis with comal tuft towards the apex, formed due to compactly placed branches and condensed nature of the internodes. Also seen are divergent and pendent branches. *B* - A stylised single stem of *Sphagnum*.

- Branching is usually lateral and fascicle of branches arises from the stem in position of fourth leaf. The fascicles are formed of two or more divergent branches and two or more pendent branches.
- At the stem apex, a crowded group of short branches arises forming a head or capitulum or comal tuft. All the branches are leafy and leaves are closely placed spirally **(Fig. 3.2)**.
- The divergent branches are short and stout and grow in a horizontal position while the pendent branches hang down close to the stem and help in water conservation **(Fig. 3.3A)**.
- The leaves on the main stem are different from the leaves of the branches. The stem leaves are small, few and spaced out, triangular to oblong or linear, while the branch leaves are numerous and crowded, oval or ovate to lanceolate.

Figure 3.3 *A&B Sphagnum. A* -The plant shows a cluster of sporophytes which are present on the female branches and are terminal in position. *B* - Portion of the apex magnified to show sporophytes borne on the stalks known as pseudopodium.

- The adult *Sphagnum* plant lacks rhizoids and absorbs water directly through the stem branches of the plant (pendent) and leaves (hyaline cells and pores).
- The stem lacks lignified tissues and does not fit the strict definition of stem, it is sometimes called a caulid. Similarly, leaves are not true leaves and are frequently called phyllidia. The stem is chlorophyllose and along with leaves forms an effective photosynthetic system.
- The rhizoids of the gametophore form a tangled mass along the stem and an efficient external conducting system at young stages.
- The sex organs are surrounded by special leaves, those around antheridia are perigonial and those around archegonium, perichaetial leaves.
- The sporophytes which are dark coloured and spherical **(Fig. 3.3B)** are rarely found in natural sites; vegetative reproduction is therefore, more common.

Anatomy

Stem

- The stem has three distinct cell types – outer cortical cells which are empty looking, hyaline, compactly arranged and thin walled. These cells may be large and store water **(Plate IV.2).**
- This is followed by an axil cylinder with two zones, the outer is formed of small, thick, pigmented prosenchymatous cells meant for support and the inner of enlarged cells which may store food **(Fig. 3.4).**

** Located in the cortical layer of divergent branches in many species are retort cells, so named because they resemble the medieval chemical flasks called retorts. These cells are enlarged more than the surrounding cortical tissue, and have a **pore**, or opening, at one end. These cells are believed to help the moss retain water under intense sunlight, and they often are home to a wide variety of small invertebrates and other microorganisms **(Plate IV.3).***

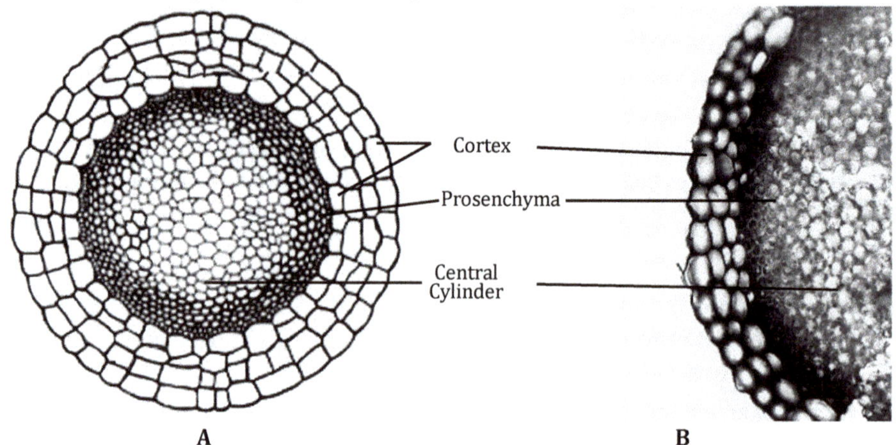

Figure 3.4*A&B Sphagnum A*-T.S. Stem with outer most cortex, middle prosenchymatous and central axial cylinder. *B*-Portion magnified to show three regions in the stem (arrow) and central axial cylinder.

Leaf

- The leaves do not have a midrib and are single celled in thickness.
- Those on the branches have two types of cells. The hyaline and chlorophyllose cells; former remain surrounded by five or six chlorophyllose cells **(Fig. 3.5).**
- The hyaline cells are large, dead with fibrous thickenings, and pores at the corners making them porous. They can absorb water and retain it also and are therefore, occasionally referred to as capillary cells.

- The chlorophyllose cells which are photosynthetic, are thinner and more elongate **(Plate IV.4).** They vary from triangular to trapezoidal or elliptic in cross-sectional view.
- The moss has an efficient means of holding water as the dried gametophore is capable of absorbing up to 20 times water than its weight. The porous cells in the leaves and also stem in certain species along with pendent branches is like a capillary wick. This compensates for the absence of rhizoids at maturity in the gametophyte.

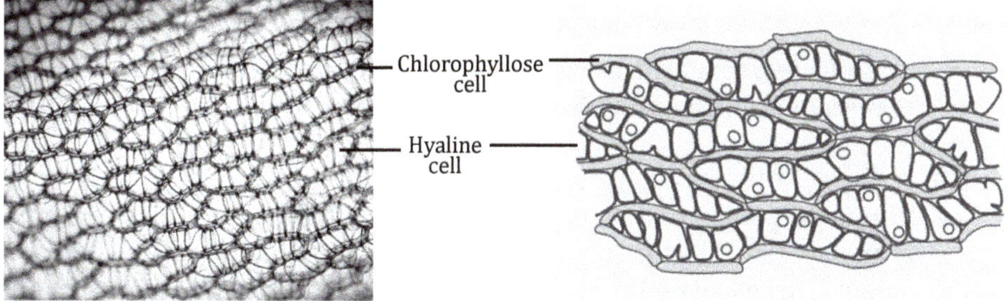

Figure 3.5 *A&B Sphagnum* W.M. Leaf. *A*-Photograph of the leaf showing two types of cells. *B*-Line diagram to show the details. The hyaline cells can be seen with strengthening bars that prevent them from collapsing. The second type of photosynthetic cell has abundant chloroplasts.

Reproduction

Vegetative Reproduction

Vegetative reproduction is through innovations, protonema, and regeneration.

- **Innovations:** It is a common method of reproduction in which special vegetative branches known as innovations take part. Occasionally one of the branches in the axillary cluster becomes robust and grows upwards. This branch is known as innovation and develops into a new plant when separated from the parent plant.
- **Multiplication of Protonemal Branches:** Any marginal cell of the primary protonema may become meristematic and form a green cellular filament which ultimately forms a new plant. Even the secondary protonema may form a leafy gametophore and bring about vegetative reproduction.
- **Regeneration:** The growth of new tissues or organs occurs to replace lost or damaged part. This is known as regeneration and is a common method of propagation.

Sexual Reproduction

The sexual branches either occur amongst the comal tuft or in the lower region of the stem. The plants may be monoecious, with both male and female branches on the same plant or they may be dioecious with separate male and female plants.

Antheridial Branch

- The sex organs are borne on special branches that develop near the apex.
- The antheridial branch has a single, stalked spherical antheridium at the base of each leaf (perigonial) in the upper third of the branch. The branches are deeply pigmented and appear red or brown.
- A mature antheridium consists of a long stalk which is eight to ten celled and a globular body enclosing spirally coiled biflagellate antherozoids.
- The antheridial jacket is unistratose or single layered.
- The antherozoids are released from the apical end of the antheridium upon dehiscence which occurs in presence of water.

Archegonial Branch

- The archegonia (one or more) are borne on short modified divergent branches that bear large leaves known as perichaetial leaves, green in colour. These leaves are protective in function, as they hold some amount of water and check dessication of the sex organs.
- Generally, a group of archegonia is seen at the apex. The first formed archegonium consumes apical cell in its formation. The archegonium which arises from the apical cell is the primary archegonium and rest of the archegonia in the group are secondary arising from apical cell derivatives.
- The mature archegonium is long and stalked structure possessing venter and a long twisted neck with six vertical rows of neck cells, enclosing eight or nine neck canal cells, a ventral canal cell and an egg cell.

Fertilization is identical with other bryophytes and requires water. But the water level generally is either high or low in the habitat of *Sphagnum* and therefore, success of sexual reproduction is low.

Sporophyte

- Fertilization leads to zygote formation which undergoes a transverse division and the lower cell gives rise to the foot which remains embedded in the gametophyte.
- The upper cell undergoes divisions resulting in two embryonic tissues, the amphithecium and endothecium; where latter forms columella and the former forms the wall and archesporium.
- The wall is multistratose. The outermost layer is epidermis which possesses non-functional stomata. The layers inner to epidermis have cells with chloroplasts. The wall or jacket encloses a dome shaped spore sac that overarches a central large columella **(Fig. 3.6)**.

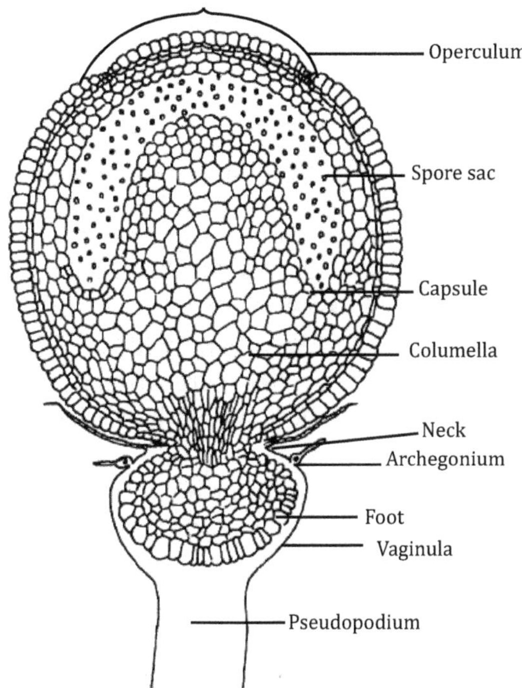

Figure 3.6 *Sphagnum* L.S. Sporophyte differentiated into a large globose capsule. The capsule shows the ill-defined lid-like operculum above the capsule which encloses a horse-shoe shaped spore sac placed above the massive columella. A small constriction represents the neck followed by a globose foot. The structure is held high by pseudopodium. Note the unfertilized archegonia around the sporophyte.

- The spores arise from the meiotic divisions of the spore mother cell, derived from the archesporium.
- Along with this development of the spores, there is formation of the annulus in the upper portion of the capsule. It is composed of small thin walled cells that form a transverse ring which delimits the operculum at the top of the capsule.
- A protective covering, calyptra is formed from the venter that covers the entire sporophyte.
- At maturity, the apex of the archegonial branch with embedded foot of the sporophyte, becomes distinct to form pseudopodium which elongates at maturity, elevating the sporophyte beyond calyptra and perichaetial leaves.
- The portion of the female stalk where the foot remains embedded is known as vaginula **(Plate IV.5)**.
- Peristome teeth are absent.

Capsule Dehiscence

- It occurs on dry sunny days when hot air enters the capsule through small spaces and wall cells and results in shrinking of the capsule. The previously spherical capsule now becomes cylindrical **(Plate VIII.3)**.
- This change in shape brings about collapse of the columella and the space created due to breakdown of collumela gets filled with hot air. With further shrinkage of the capsule, this trapped air comes under pressure which is about 5atm. The pressure is passed on to the operculum which is attached to the capsule by annulus **(Box 3.1)**.
- The compressed air inside capsule tries to escape and as it does so, it pierces through the spore sac. The spore sac tears, leading to throwing off the operculum and flicking spores violently as a bullet from the gun **(Plate IV.6)**.
- The spores are shot into the air to far-off distances where they germinate to form a thalloid protonema thus, starting the gametophytic phase of life cycle **(Fig. 3.7)**.

Figure 3.7 *Sphagnum* W.M. showing thalloid protonema.

Box 3.1

The maximum discharge speed of spore dispersal measured is reported 3·6 m s^{-1} and the maximum height to which they reach is 20cm (average: 15cm) above the capsule (Sundberg, 2010). Peat mosses possess a unique and well-known mechanism of violent spore discharge, mentioned first by Linnaeus (Linne, 1771, p. 506: 'Sphagn. palustre. Antheræ dissiliunt cum sono et crepitu. Angerstein.'), in which spores are ejected from the spore capsule with a sharp noise audible at a distance of several metres, termed the 'air-gun mechanism' (Nawaschin, 1897; Ingold, 1965).

The pricking experiments demonstrate that the air-gun notion for explosive spore discharge in *Sphagnum* is inaccurate; differential shrinkage of the capsule walls causes popping off the rigid operculum. The absence of evidence for a potassium-regulating mechanism in the stomatal guard cells and their gradual collapse before spore discharge indicates that their sole role is facilitation of sporophyte desiccation that ultimately leads to capsule dehiscence (Jeffrey *et al.,* 2009). The nineteenth century air-gun explanation for explosive spore discharge in *Sphagnum* has never been tested experimentally. Similarly, the function of the numerous **stomata** ubiquitous in the capsule wall has never been investigated. Both intact and pricked *Sphagnum* capsules, that were allowed to dry out, all dehisced over eight to twelve hours period during which time the **stomata guard cells** gradually collapsed and their potassium content, measured by X-ray microanalysis in a cryoscanning electron microscope, gradually increased. By contrast, guard cell potassium fell in water-stressed *Arabidopsis.* Functional data on *Sphagnum,* when considered in relation to bryophyte phylogeny, suggest the possibility that **stomata** first appeared in land plants as structures that facilitated sporophyte drying out before spore discharge and only subsequently acquired their role in the regulation of gaseous exchange.

The life cycle of the bog moss is given in **Figure 3.8.**

Unique features of *Sphagnum*:

- Fascicles of branches with divergent and pendent types.
- Leaves with hyaline and chlorophyllose cells.
- Thalloid protonema similar to leafy liverworts.
- Position of sex organs similar to leafy liverworts.
- Explosive dehiscence of capsule and shot gun mechanism of spore dispersal.

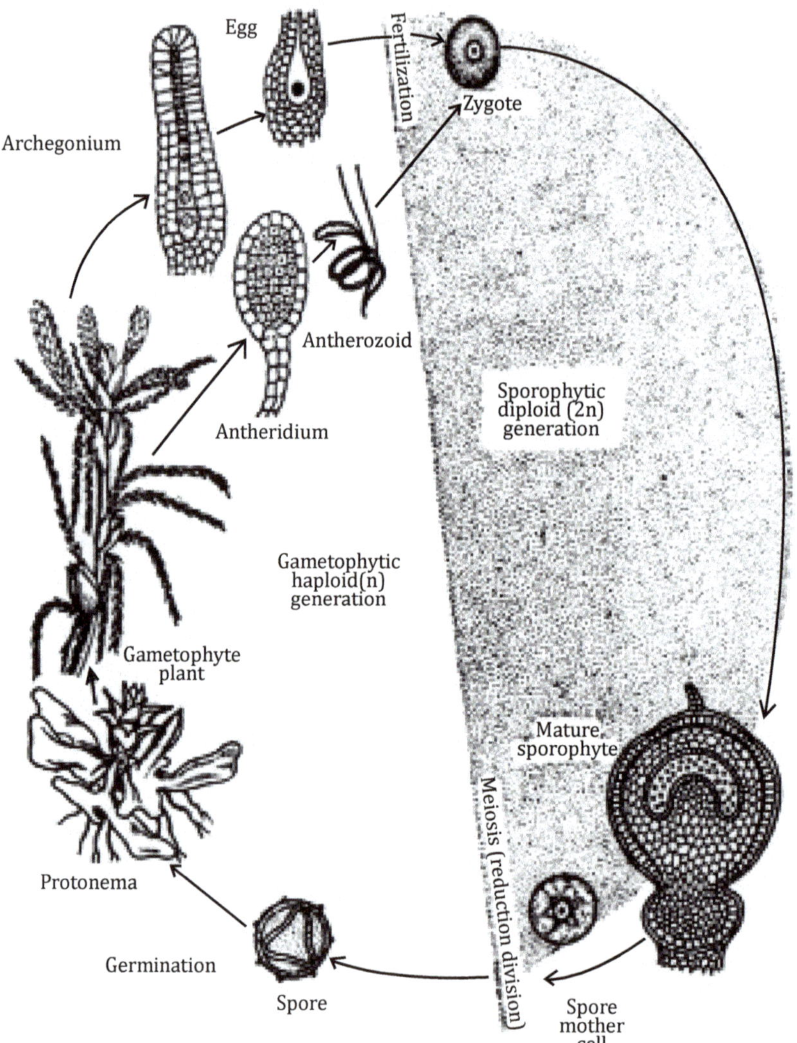

Figure 3.8 *Sphagnum* Life cycle with significant stages.

- Absence of rhizoids at maturity but ability to absorb tremendous amount of water through pendent branches and leaves.

Evolutionary Trends (As considered by early schools that supported progressive sterilization of tissues)

The sporophyte in mosses is generally complex and differentiated into foot, seta and capsule. The seta in more advanced mosses has a conducting strand with thick walled cells. The capsule wall has stomata leading to chlorophyllous tissue with intercellular spaces, thus creating an efficient gas exchange system. The capsule has an operculum and peristome, a highly efficient region meant for dehiscence and spore dispersal.

Sphagnum resembles true mosses only in a few characters and it also bears resemblance to Jungermanniales. However, *Sphagnum* is enigmatic moss because most of its features are unique, not encountered in any other plant. Hence *Sphagnum* is a referred to as synthetic genus **(Table 3.2)**.

Table 3.2: Characters of *Sphagnum* comparable to various groups of bryophytes.

Marchantiopsida	Bryopsida	Anthocerotopsida	*Sphagnum*
Protonema—flat, plate-like (similar to acrogynous jungermanniales	Leafy gametophore	No apical growth in sporophyte	Absence of rhizoids in adult gametophyte
Resembles *Porella* in position and dehiscence of antheridium and position and development of archegonium	Rhizoids—multicellular with oblique septum	Origin of archesporium from amphithecium	Branches fasiculate Absence of costa in leaves Leaves with hyaline and chlorophyllose cells
	Development of antheridium.	Origin of columella from endothecium	Stem cells with pores
	Absence of elaters	Capsule wall is photosynthetic	Antiseptic properties due to acidic substances
	Dehiscence of capsule at operculum	Large, bulbous foot, seta indistinct	Peculiar water absorption and conduction

Ecological Value of Peat Moss

- Peatlands occupy nearly 3% of the land surface and store 25% of world's soil carbon of recalcitrant peat (Yu *et al.,* 2010).
- *Sphagnum* represents an unparalleled model system for ecological and evolutionary genomic resources.
- It plays a major role in peat land formation, a prime example of ecosystem engineering. It involves the accumulation of peat that facilitates its own growth while making the surrounding hostile for vascular plants (van Breeman, 1995).
- Hence, *Sphagnum* has a remarkable ability to create and then uniquely thrive in nutrient-poor, acidic and water-logged conditions.
- The morphological and physiological traits help it in creating its micro habitat. For example, its ability to store and transport water is controlled largely by – branching architecture, leaf size and their arrangement on branches and hyaline cells. Hummock- forming species that grow approximately more than 30cm above the water table, have small close-set leaves forming numerous interconnected small capillary species.

Spreading branches allow lateral movement of water through capillary continuum while numerous pendent branches adpressed to the stem form vertical water transport. Further, the dead hyaline cells in leaves and outer cortex of stems with retort cells and branches act as water storage sites.

- When the plants die, pore size reduces in leaves, making it and subsequently the habitat, water-logged.
- The mechanism by which it inhibits fungal and microbial decomposition is attributed to cell wall polysaccharide hydrolysis and release of 'sphagan' in the process. Soluble phenolics leached from *Sphagnum* thalli also play a role in slow decomposition.
- Hyaline cells not only play a vital function as water storage organ but also create a novel and safe habitat for diverse microflora spanning all domains of life (Weston *et al.*, 2017).
- Diverse communities of microorganisms are intimately associated with *Sphagnum* and these contribute to the health and productivity of the plant.
- The violent spore release in *Sphagnum* clearly works as a very efficient way to increase flight (dispersal). This must be important because peat mosses do not possess as in other mosses, other mechanisms aiding passive spore liberation such as long and thin setae, hanging capsules or peristome teeth.
- Peat mosses are literally in a class of their own (these are the only mosses with direct and substantial economic value and peat mosses have been used by people for centuries (Turner, 1993). Many of their uses derive from the extra ordinary absorptive capacity of peat mosses (as in bandages, diapers, women hygiene products, which in turn derives from the unique morphology of *Sphagnum*).
- The peatlands constitute an important reservoir for global carbon and currently function as carbon sink (Gorham, 1991). *Sphagnum*-dominated peatlands also have a profound effect on gas fluxes that are determinants of global climate (*e.g.*, CO, N_2O, H_2S and DMS).
- *Sphagnum* is distinct from all other mosses in numerous aspects of both sporophyte and gametophyte morphology.

Ecologically it helps in land reclamation. When the plant grows on lakes, the intertwined moss gives appearance of solid surface and forms quacking bogs. Other hydrophytic plants grow on it and then slowly after death and decay of these plants and also of the bog moss, debris is formed. This process continues and the surface gets raised up. On this dead moss debris, subsequently a different type of mesophytic flora appears.

Economic Value

The peat formed over several years due to partial decomposition of the dead plant material is used as a fuel. It played an important role in World War II during which it was used to dress wounds of the soldiers due to antiseptic properties of the thallus and now is used to prepare toiletteries **(Plate IV.7A).** A liquid-absorbent article having a high absorption capacity and a short fluid penetration time has been prepared. The liquid-absorbent article comprises *Sphagnum* moss material containing an effective amount of cross-linked cellulosic fibres. The liquid-absorbent article is well-suited for use as an absorbent component of a disposable absorbent product, such as a sanitary napkin **(Plate IV.7B),** a diaper, an incontinence pad, an adult brief, a wound dressing, a nursing pad, a tampon pledger, or as desiccant for packaging materials to keep goods dry during shipping or storage. The invention also extends to a novel method for manufacturing the liquid-absorbent article (https://patents.google.com/patent/US5718697A/en). As the plant holds lot of water, it is widely used in horticulture to wrap seedlings while transporting them from one place to another.

Class Polytrichopsida Doweld (after Goffinet *et al.*, 2008)

- Mosses in the class Polytrichopsida are often considered as pioneer plants (after Goffinet *et al.,* 2008). They exhibit a wide distribution. These mosses occupy acidic, exposed and nutrient-poor soils.
- The acrocarpous species display great diversity ranging from the miniature plants of *Pogonatum pensilvanicum* with its highly reduced leaves, to the large gametophyte of *Dawsonia superba*, which can reach heights as tall as 50cm.
- Not only are the leaves of this class costate, but they also possess photosynthetic lamellae forming a 'pseudo-mesophyll' (Smith, 1971). Variations in the height, number and morphology of lamellae along with the morphology of the lamina are often used as species specific character.
- The vascularization in Polytrichopsida is well developed.
- The peristome is nematodontous. Mosses characterized by nematodontous peristomes form a grade paraphyletic to the arthrodontous taxa.

Order Polytrichales M. Fleisch

Includes highly abundant taxa that may dominate certain communities, especially in the northern boreal zone (*e.g.,* certain species of *Polytrichum* Hedw.).

- Tall, perennial gametophores have narrow leaves. These leaves are with lamellae and distinct midrib or costa.
- The capsule is erect to horizontal.
- The peristome teeth are 32 or 64 arranged in single circle or ring and are of nematodontous type, made up of overlapping whole dead cells.

- The columella expands into thin epiphragm that covers the mouth of the capsule.
- Calyptra may be smooth or hairy.

Family Polytrichaceae Schwagr

The Polytrichaceae known to include 'Aloe mosses', is widely distributed throughout the world, and diversity is highest in South East Asia and South America with 7 genera and 14 species in Australia. Plants grow in tufts, scattered or gregarious, on soil, humus or peat, rarely on rock. The family is an important component of the pioneer plant communities of disturbed soil, and many of the species are light-tolerant and xerophytic.

Polytrichum – The Hair Cap Moss

Habitat and Distribution

The genus has about 100 species, common being *P. juniperinum*, *P. xanthopilum* and *P. alpinum*. They inhabit varied regions.

Gametophyte

The spore upon germination gives rise to a protonema with a horizontal and an upright system. The horizontal underground rhizome gives rise to the sparingly branched leafy shoot while a complex system of rhizoids arises from the hyaline portion of the protonema.

The plants are dioecious with separate male and female plants. Upon fertilization, a zygote starts the sporophytic generation. The sporophyte with a capsule, seta and foot, is a complex structure. Spore dispersal is complex and occurs in a highly advanced manner. The calyptra usually appears as a hairy cap and is a distinguishing feature of the moss which is therefore, commonly known as – HAIR CAP MOSS.

Morphology

The gametophore is differentiated into a horizontal rhizomatous system and an aerial upright leafy axis that bears sporophyte **(Fig. 3.9A; Plate V.1).**The rhizomatous region gives rise to rhizoids while the aerial axis has large leaves which are clustered to form heads during reproductive period. Gametophyte bears long cylindrical sporophyte distinguished by hairy calyptra **(Fig. 3.9B; Plate V.2).**

Leafy Axis

- The plants are much branched and spreading.
- Leaves on the lower part of the stem are brown and scaly and with no distinct spiral arrangement while those on the upper portion of the axis are with a distinct 3/8 spiral arrangement **(Plate V.3).**
- Leaves are long to lanceolate, costate and possess membranous expanded sheaths **(Fig. 3.9C).**

- The sheath clasps the stem and acts as an external capillary system for water conduction.
- This sheath shows transition into a blade and the transition point has the hinge tissue **(Plate V.4).**
- Distinct midrib or costa that runs through the leaf, is broad at the base tapering towards the tip.
- The wing is coarsely toothed.

Rhizoids

- Arise from the horizontal portion of the gametophore as long thread-like structures.
- The rhizoids are typical of mosses branched with oblique septum **(Plate V.5).**
- The secondary and tertiary branching of rhizoids is seen with intertwining of the rhizoidal branches.
- This results in formation of thick twisted cable-like wicks which assist in external capillary action of water **(Fig. 3.9D).**
- Thus, rhizoids are not only involved in anchorage but also assist water conduction.

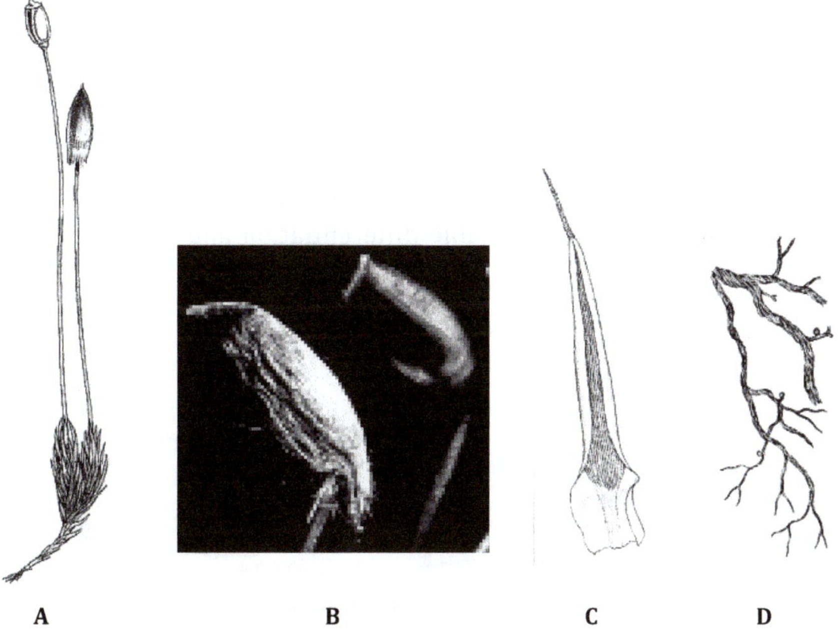

Figure 3.9*A-D Polytrichum. A* - Leafy gametophyte bearing sporophyte which represents the diploid phase. *B* - Note the long seta and hairy calyptra. *C* - Leaf shows a distinct hinge and a midrib. *D* - Multicellular rhizoids are branched and show oblique septa.

Anatomy

Leaf

- At the limb portion, several celled thick midrib is seen which tapers towards the rudimentary wings **(Plate V.6).**
- The anatomy of the leaves is structurally complex. The abaxial part of the leaf consists of large cells with thick walls, the hydroids, which help in the transport of water, and of leptoids, which are efficient in the transport of sugars (THOMAS *et al.,* 1990). The closely arranged longitudinal rows of one cell layer wide chlorophyll-containing, lamellae on the adaxial side of the leaf have the same function as the mesophyll in higher plants **(Fig. 3.10A)** (PAOLILLO and REIGHARD 1967, THOMAS *et al.,* 1996).
- The lamellae are formed of parallelly arranged chlorophyll containing cells with five to eight cells per lamella **(Fig. 3.10B).**
- The notched terminal cell is larger, in close proximity with the neighbour and results in ill-defined upper epidermis **(Plate V.6B).**
- The lowermost is the epidermis with outer thickened walls.
- The central tissue has large thin walled cells among which are scattered small steroidal cells.
- The photosynthetic lamellae compensate for the reduction in size of the wing as far as photosynthetic efficiency is concerned.

**The sheath lacks lamellae and the costa is made up of thick walled supportive cells with scattered hydroids and leptoids that help in water conduction.*

Stem

- The outline of the rhizome and also stem is not circular but appears shallowly ridged with several leaves also seen in the section **(Plate V.8).**
- Stem shows remarkable differentiation and the outermost layer is

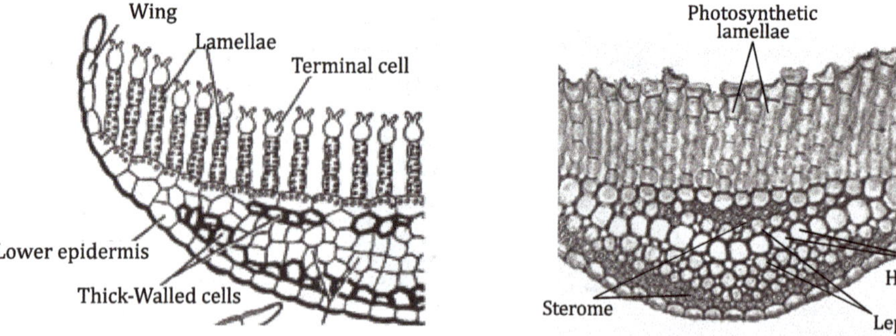

Figure 3.10*A&B Polytrichum* V.S. Leaf *A*-Line diagram. The terminal cell of the photosynthetic lamella is bifurcated. *B*-Cellular details - Taking a closer look at the costa, the water conducting hydroids visible as the larger row of cells are near the middle. These have large, empty spaces inside to provide ample space for water to flow. The cells above and below form the food-conducting leptoids and the surrounding cells with the thickened cell walls,having supportive function are called stereids.

epidermis. In case it is the subterranean region, the epidermis is non chlorophyllous and shows rhizoids arise from epidermis. In this case, the tissues are also not as distinct as they are in aerial stem **(Fig. 3.11A).**

- Inner to epidermis, is a large multilayered cortex with two or three layers of parenchyma interrupted at three places by hypodermal strands situated opposite three ridges.
- These strands are made up of steroidal cells which are thick walled and are meant for support. The tissue formed by steroidal cells is called sterome.
- Inwards, the hypodermal strands extend into larger, thick walled cells and together both form radial strands **(Fig. 3.11B&C).**
- The radial strands at the point of insertion in the central cylinder possess six to eight cells rich in proteins and are known as leptoids, the tissue called leptome and is considered analogous to sieve elements but lack p-proteins.
- The leptome is demarcated from the central cylinder by starch containing cells known as amylome.
- The central portion is made up of mainly thick walled cells, among which are present empty looking water conducting cells, the hydroids. The tissue formed by these hydroids is called hydrome. The hydroids are analogous to tracheids but totally lack lignin **(Fig. 3.11; Plate V.8).**

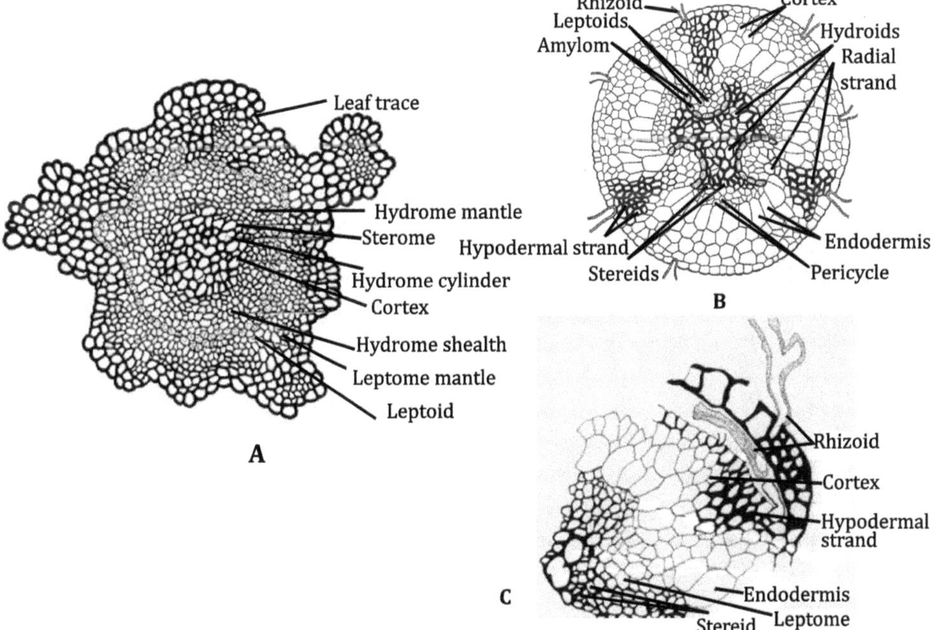

Figure 3.11 ***A&C Polytrichum. A*-T.S. Rhizome with hydrome in the centre that helps in water conduction and is equivalent of xylem while the leptome represents the phloem of the higher plants. *B*-T.S. Stem shows distinct radial strands along with the hydrome, sterome and leptome. *C* - Shows details of one of the radial strands.**

Reproduction

Vegetative propagation occurs as in other mosses.

Sexual Reproduction

Plants are dioecious. The sex organs are organized in 'head-like' structures which have antheridia/archegonia interspersed with perigonial/perichaetial leaves **(Fig. 3.12A).**

Antheridial Head

- Antheridia are borne in clusters at the apex of male branch.
- They are covered by involucre of perigonial leaves **(Fig. 3.12A)** which are brightly coloured **(Plate V.9).**
- The apical cell is not consumed in antheridia formation, hence the vegetative growth of the male branch is not stopped.
- The mature antheridium is a club-shaped structure, borne on a short stalk **(Fig. 3.12B)**.
- Interspersed with antheridia are paraphyses, the sterile hair-like structure.

Archegonial Head

- The female head is essentially like the male with a difference that the apical cell is consumed in archegonial formation and the number of archegonia is also less **(Fig. 3.12C).** The leaves associated with female sex organs are perichaetial leaves. The archegonia have a long neck and a swollen venter **(Fig. 3.12D).**

Fertilization results in starting up of a new generation, the sporophyte.

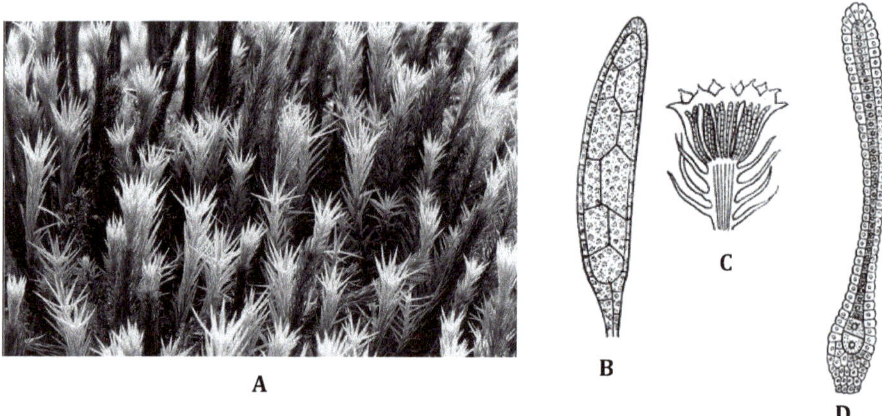

Figure 3.12*A-D Polytrichum* Sex organs. *A* - Reproductive heads have resemblance with flowers of higher plants. The perigonial leaves are compactly arranged. *B* -The antheridia are club-shaped with a distinct stalk. *C* -The archegonial head. *D* - The enlarged archegonium with a long neck.

Sporophyte

- The mature sporophyte is differentiated into a foot, a long seta and a capsule.
- Parenchymatous foot is deep seated in the archegonial branch.
- Seta which carries the capsule atop, is slender. The architecture of the seta shows epidermis as the outermost layer, inside, it leads to sclerenchymatous, and then a parenchymatous region with cells that enclose intercellular spaces. These cells are rich in chlorophyll. Innermost region is occupied by a central strand meant for support.
- The base of the capsule that merges with seta is a modified region known as apophysis. The apophysis has functional stomata and chlorophyllous tissue.
- The capsule is angular and slender **(Fig. 3.13A).**

Capsule

- As the capsule is angular, the outline of the transverse section is not circular **(Fig. 3.13B).**
- The epidermis forms the outermost layer of the capsule wall.
- This is followed by a layer or two of parenchymatous cells that lead to air space (outer) traversed by radially situated assimilatory filaments which connect the wall to the spore sac.
- There is also present an inner air space that connects spore sac to central columella which is also traversed by radially situated assimilatory filaments **(Fig. 3.13B)**.
- The spore sac wall is two layered and encloses only the spores.
- The apical region of the capsule has operculum that appears as a lid with a long beak **(Plate V.10).**
- Peristome is formed of 32 or 64 pyramidal teeth which are non-ornamented; connected upwards to membranous epiphragm. The epiphragm is large and peristome is small which does not respond to changes in moisture as a result of which they do not show movements **(Fig. 3.13C).**

Dehiscence and Spore Dispersal

- At maturity seta elongates and the calyptra is ruptured leading to removal of operculum.
- The spores are dispersed when the operculum falls off under dry conditions. This is facilitated by peristome.
- When moisture is available, epiphragm remains loose and teeth are not stretched. There are no spaces between teeth. However, when dry

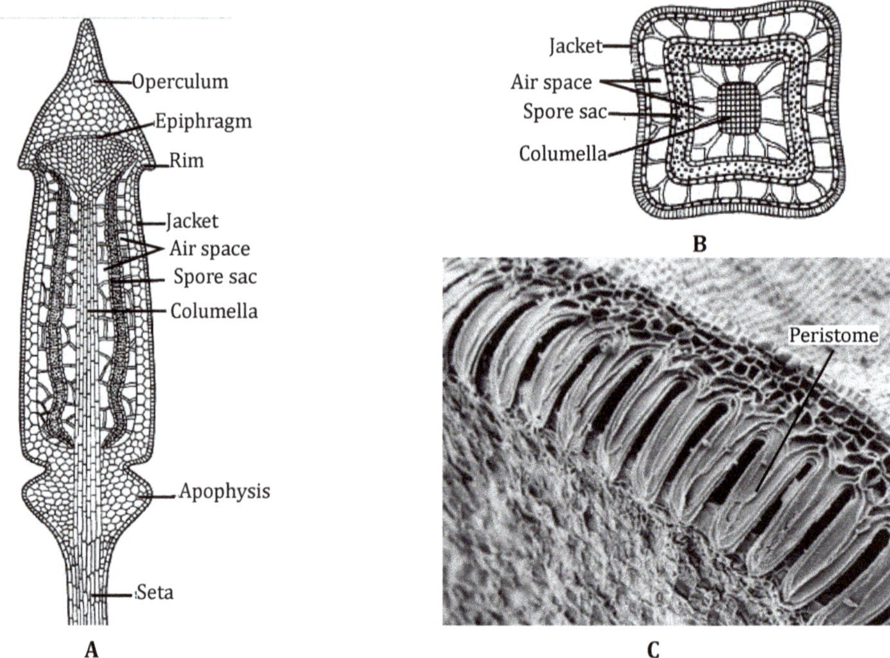

Figure 3.13*A-C Polytrichum*. *A*-L.S. Sporophyte differentiated into seta, apophysis, theca and operculum. Note the fertile tissue is restricted to a small narrow spore sac. *B*-T.S. Capsule shows wall, air spaces, spore sac and central columella. *C*-The peristome, non ornamented is arranged in a single ring.

conditions are available, epiphragm loses water and is under stress, teeth get stretched and spores are sifted through the gaps formed in the margin of the epiphragm **(Plate V.11).** This adjustment allows only a few spores to be released at a time and is known as censor mechanism **(Plate VIII.4A &B).**

Protonema

Spore germinates to form a protonema with two systems; the rhizoidal and the upright, from latter arises erect leafy axis of the gametophore. This starts gametophytic phase and life cycle of the moss is complete **(Fig. 3.14).**

Class Bryopsida – Arthrodontous Mosses (after Goffinet *et al.*, 2008)

Bryopsida is a species-rich group with mosses that show advanced characters and adaptability to explore various habitat conditions. They are marked by distinct and articulate peristome.

Subclass Funariidae Ochyra

- The leaves bear a distinct midrib and are several celled thick.
- Archesporium, the fertile region of the capsule, arises from the outer layer of endothecium, one of the two embryonic layers.

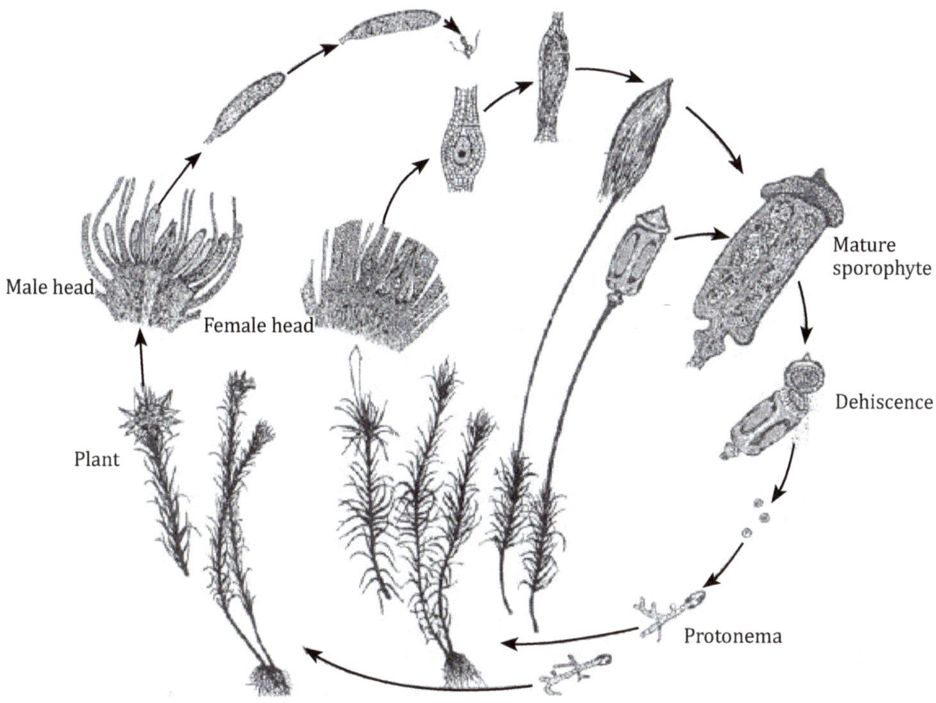

Figure 3.14 *Polytrichum* Life cycle of a dioecious moss.

- The spore sac remains connected to the capsule wall by trabeculae.
- Seta is well developed and performs the function of elevating capsule above protective layers as well as facilitating spore dispersal.
- Capsule shows peristome around the mouth of spore cavity. The peristome teeth are ornamented.
- Dehiscence of capsule and spore dispersal are highly elaborate mechanisms.
- Protonemal stage is distinct and filamentous.

Order Funariales M. Fleisch.

- Leaves are small and remain clustered especially in sexual stages, due to which the plant gets appearance of a head.
- The capsule is broad and has a lid or operculum without a beak.
- Peristome comprises two rings, outer and inner peristome or exo-and endostome .
- Peristome is of arthrodontous type *i.e.,* a peristome in which the teeth are structured by articulated cell wall.

Family Funariaceae Schwagr.

- Leaves are distinctly costate usually one celled thick at the wing region.
- Pyriform capsule is placed in a manner that it is nodding at maturity.
- Seta is long and twisted.

Funaria – The Cord Moss

Habitat and Distribution

The genus *Funaria* commonly known as cord moss or green moss, is cosmopolitan in nature. It forms velvety tufts on moist ground, rocks, tree-trunks and under shade. *Funaria hygrometrica* is one of the most common, weedy, and widely distributed mosses in the world. Its distribution closely parallels that of *Bryum argenteum*. It appears along with *Marchantia polymorpha* in the areas which have been subjected to fire. The genus has about 117 species with wide distribution throughout the world; eight species occur in India. The best known species is *F. hygrometrica*.

Gametophyte

The spore germinates to form a protonema with two systems, caulonema and chloronema. The latter develops buds which give rise to leafy gametophore about 1-3cm long, with erect and branched stem bearing spirally arranged leaves. The upper leaves are large and lower leaves are small and crowded. The plant remains attached to the substratum by rhizoids, typical of mosses.

During the sexual phase, plant bears antheridia and archegonia. The plants are autoecious monoecious with separate male and female branches on the same plant. The sex organs are interspersed with paraphyses forming 'heads'.

Funaria is protandrous *i.e.*, male sex organs mature first making the plant depend on cross fertilization for sexual reproduction. Water is essential for fertilization. When the surface of moss plant is wet, the mature antheridia absorb water trapped between perigonial leaves and paraphyses and burst releasing the male gametes (antherozoids) through apical portion. The male gametes swim towards archegonia. In each mature archegonium, neck canal cells and ventral canal cell degenerate to form mucilage which swells up and opens the archegonium to create a passage up to egg. The mucilage contains sucrose which attracts the male gametes (chemotactic).

After fertilization, a zygote is formed which undergoes divisions to form an embryo with two growing points. The two embryonic layers endothecium and amphithecium arise which give rise to different types of tissues in the different regions of the sporophyte. The sporophyte is highly complex and divided into upper part with operculum and peristome, central part with theca proper and lower part or the apophysis. The sporophyte remains attached to the gametophyte through a small foot which remains connected to the capsule by a slender seta. The capsule is a highly differentiated structure and bears haploid spores at maturity. The capsule dehisces by a very efficient mechanism and spores are released by sieve mechanism. The spores germinate to form a filamentous protonema and the leafy gametophore gets established.

Morphology

- Gametophore is differentiated into stem and leaves. It is found growing in tufts or clusters **(Fig. 3.15A, Plate VI.1).**
- Rhizoids arise from the base of the gametophore and anchor it into the soil **(Plate VI.2).** They also have a role in absorption of water.
- The leaves are small, oval, sessile and green **(Plate VI.3).** The leaves borne on prostrate branches and on lower portion of an erect branches are pale to non-green and scale-like, whereas the leaves on upper portion of an erect branches are green and are larger in size. These are called foliage leaves and are spirally arranged on the stem **(Fig. 3.15B).** They surround sex organs forming 'heads' at the apices of special branches.

Anatomy

Stem

- The outermost layer is epidermis with cells that possess chloroplasts. Below the epidermis, there is multilayered parenchyma, forming cortex.

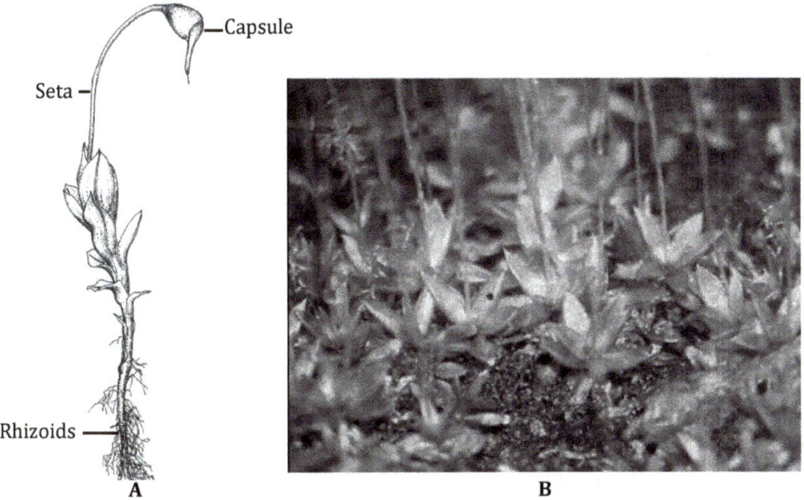

Figure 3.15*A&B Funaria. A*-Plant with closely placed leaves on the lower part of the stem that bears the diploid phase, well differentiated into foot, seta and capsule. The lower portion of the stem shows fine thread-like rhizoids. *B*-The plant grows in clumps or clusters and several leafy gametophores arise close to each other.

- In the centre, there is compact zone of cells which are empty looking, without protoplast and form a central cylinder which functions as a conducting tissue **(Fig. 3.16).**

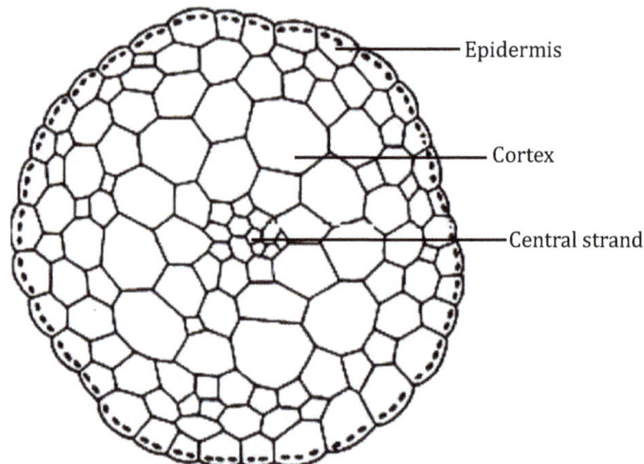

Figure 3.16 *Funaria* T.S.Stem showing single layered epidermis followed by a large cortex that encloses a simple cylinder.

Leaves

- The leaf is differentiated into lamina or wing and a distinct midrib or costa **(Fig. 3.17A, Plate VI.4).**
- The leaf lamina or the wing consists of single layered parenchyma cells, rich in chloroplasts. Several celled midrib is present. The central core is formed of thin walled cells supported by thick walled cells on either side **(Fig. 3.17B, Plate VI.5).**

Reproduction

Vegetative Propagation

Alternation of generations through sexual mode is not always obligatory as the moss frequently shows alternative reproduction modes. Frequently, the gametophyte undergoes vegetative propagation to form a succession of the gametophyte generations before the sporophyte generation develops. Vegetative propagation is a method by which a part of the gametophyte separates or specializes to form a new individual.

Fragmentation of Primary Protonema

Accidental damage, or death of intercalary segments (also known as separating cells) cause fragmentation of primary protonema, with each fragment giving rise to a leafy gametophore.

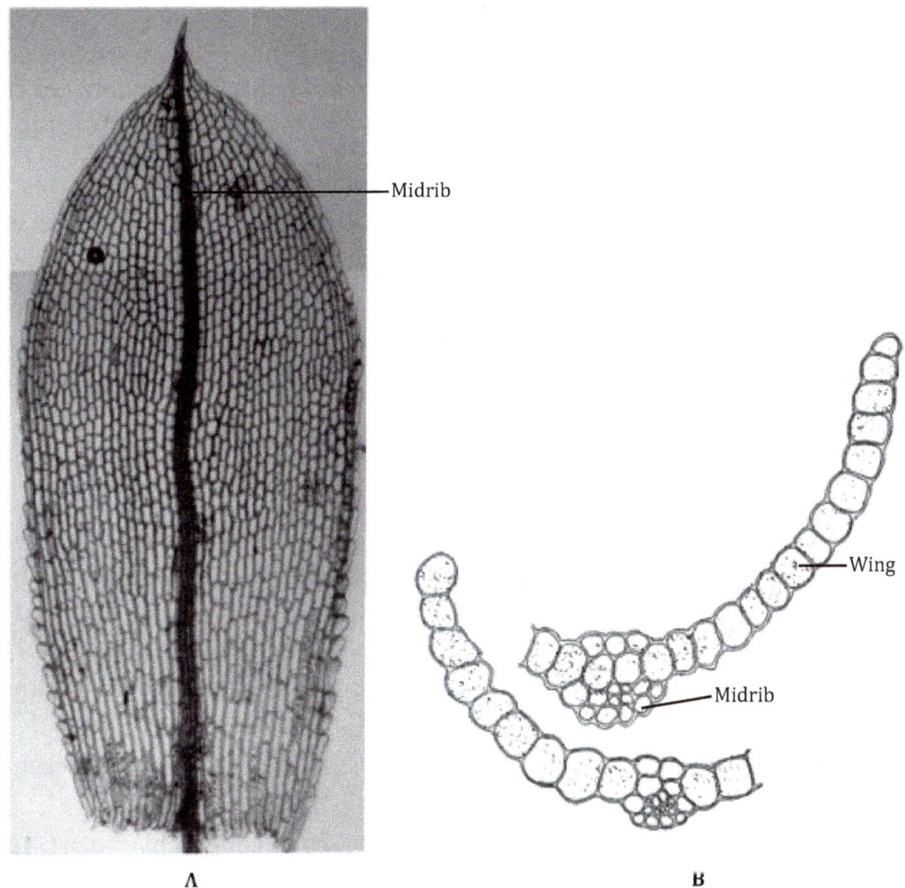

Figure 3.17*A&B Funaria. A* - W.M. leaf with a midrib or costa that runs from the base to the tip. *B* - Cells of the wing are rich in chloroplasts while the midrib shows thick walled cells for conduction.

By Secondary Protonema

Any cell of the injured part of the gametophore may give rise to secondary protonema which develops lateral bud giving rise to a leafy gametophore.

By Bulbils (Tubers)

These are perennating structures, developed on rhizoids under unfavourable conditions. On return of suitable conditions, each of them produces a protonema.

By Gemmae

Occasionally, the apices of protonemal branches develop gemmae which form gametophore under favourable conditions.

Apospory

This is the vegetative propagation of sporophyte in which any somatic cell produces leafy gametophore without spore formation. Such gametophores are diploid (2n).

Sexual Reproduction

Adult gametophore is monoecious and autoecious *i.e.,* male and female sex organs develop at the tips of separate shoots of the same gametophore.

Antheridial 'Head'

- The main axis acts as a male shoot. The tip of male shoot has a convex disc or receptacle on which a cluster of club-shaped antheridia are intermingled with paraphysis arises **(Plate VI.6).**
- The antheridia of all stages lie intermixed without any particular age-wise segregation of maturity.
- The leaves on the lower part of the male shoot are small and scattered but towards tip are crowded together to form a rosette or a 'head' resembling a 'flower'.
- The 'head' is surrounded by rosette of perigonial leaves that are very stiff and form a fairly robust, cup-like structure. The antheridia occur within the whorl of perigonial leaves **(Fig. 3.18).**
- Mature antheridium has a short massive stalk and a club-shaped jacketed body.
- The jacket is single layered with each cell containing a chloroplast. The carotenoids impart orange colour to the jacket. A few cells towards the apex enlarge and their walls become thick. These are the opercular cells which help in dehiscence and release of antherozoids at maturity.
- Inside the jacket, a mass of androcyte mother cells is present, each of which divides diagonally into two androcytes. Each androcyte later develops into a biflagellate antherozoid.
- Once the antherozoids are mature, there is a two-phase antherozoid release following the disintegration of the opercular cells of the antheridium.
- The antheridia lack any fluid at the base and the first ejection phase is governed by contractions of the antheridial walls.
- The second is the slow-release phase, during which the remaining antherozoids are not pushed out by water uptake of internal antheridial fluid but they are pulled out because of the cohesiveness of the gamete mass and the swelling of the extruded portion (up to five times its volume when contained in the antheridium).
- The gamete mass is discharged into the water-filled, reddish cups from where the rain drops splash these cells further away. Fats within the

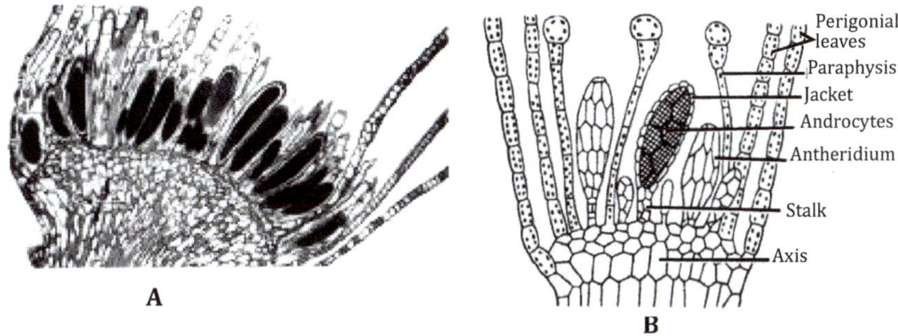

Figure 3.18 *A&B Funaria* L.S. Antheridial head. *A* - Shows club-shaped antheridia intermixed with paraphyses and perigonial leaves. *B* - Line diagram to show the structures that comprise the male head. The rounded terminal cell of the paraphysis is indicated to play a role in keeping the antheridia moist at maturity.

gamete masses ensure that they break up, making it easier for a falling drop to dislodge some of the antherozoids.

- The same splash-cup mechanism is used by various other mosses *e.g., Polytrichum juniperinum*, though the splash cups are not so strikingly coloured **(Plate VIII.2)**.

Archegonial 'Head'

- The female shoot arises from the base of the male shoot and called as archegonial branch.
- The female shoot is also leafy but unlike male shoot, leaves are not much different from the vegetative branch.
- The female branch has a limited growth as the apical cell is used up in archegonium formation.
- A few archegonia arise in the archegonial 'head' which also has associated perichaetial leaves **(Fig. 3.19A, Plate VI.7)**.
- Each archegonium has a long and a massive stalk, flask-shaped venter and a long neck **(Plate VI.8)**. Venter encloses a basal egg cell and upper smaller ventral canal cell.
- Neck consists of six or more neck canal cells, and is made of six rows of cells and is slightly obliquely placed.
- The jacket is double layered near the stalk, becoming single layered above **(Fig. 3.19B)**.

Sporophyte

- Is a complex structure differentiated into foot, seta and a capsule **(Fig. 3.20)**.

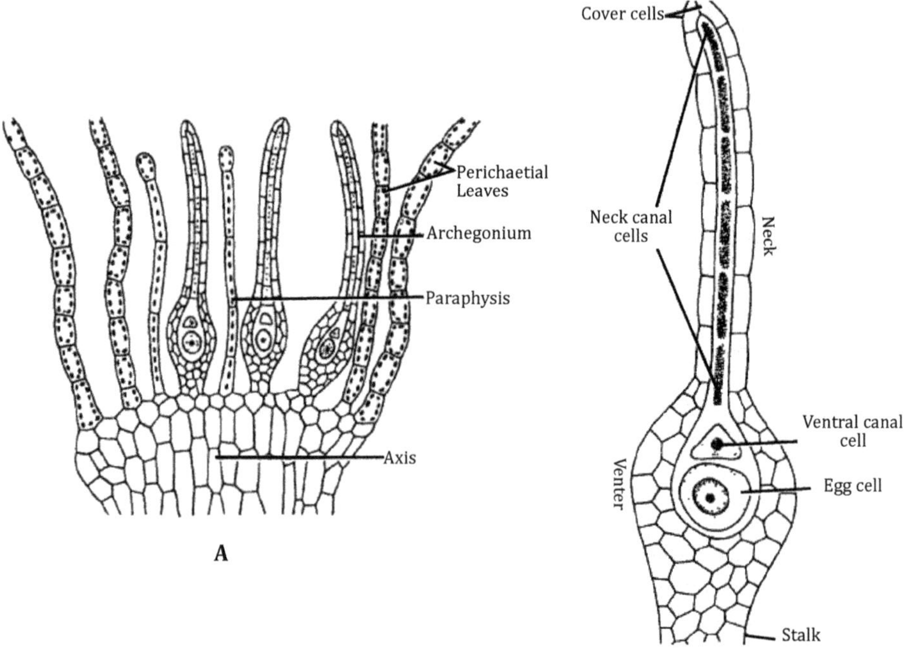

Figure 3.19*A&B* *Funaria* L.S. Archegonial head. *A* - The head shows perichaetial leaves and flask-shaped slender archegonia. *B* - Single archegonium to show a long neck, swollen venter and a stalk.

- The foot is a dagger-like or conical structure embedded in the apex of archegonial branch and is assigned the function of absorbing water and inorganic nutrients besides, anchoring the sporophyte in the gametophyte.
- Seta is a long and twisted stalk with a pear-shaped, obliquely placed capsule at its tip.
- Capsule has three distinct regions - basal apophysis, central theca and terminal operculum **(Plate VI.9).**
- Annulus is a ring of cells that separate operculum and theca, the fertile region **(Fig. 3.21, Plate VI.10).**
- As the sporophyte develops, the venter divides along with it and forms a protective covering called calyptra around the venter, which ruptures at maturity and remains like a cap on the capsule. Calyptra is haploid because it develops from venter wall.

Fertilization mediated by water occurs leading to zygote formation. The zygote forms an embryo which gives rise to two embryonic layers – amphithecium and endothecium. In fertile region, amphithecium gives rise to wall of the theca while endothecium forms archesporium and columella and inner spore sac. In apical portion, amphithecium forms operculum and peristome while endothecium forms tissue continuous to columella. In apophysis region amphithecium forms wall layers and endothecium, the conducting strand.

Operculum and Peristome

- Is the region associated with dehiscence of the capsule and dispersal of spores

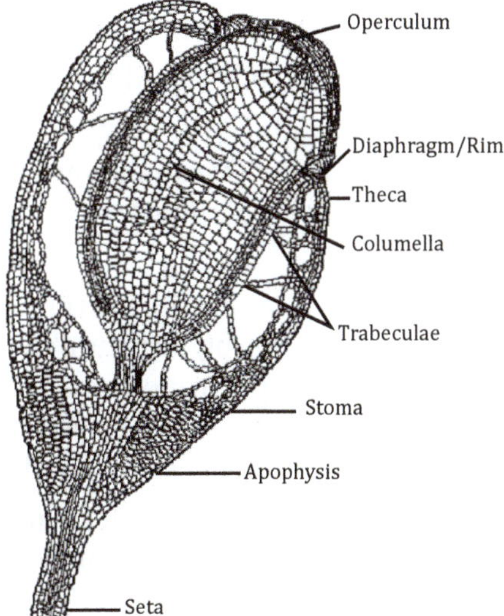

Figure 3.20 *Funaria*. Diagrammatic sketch of sporophyte, a highly advanced and a differentiated structure. Note greater sterilility in form of trabeculae, multistratose wall and a central columella.

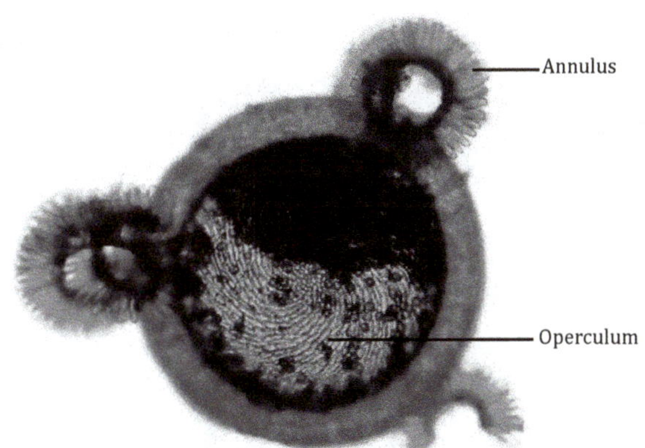

Figure 3.21 *Funaria hygrometrica var. hygrometrica*, photomicrograph of dehisced operculum with spiral pattern of cells and with adherent segments of revoluble annulus (*Source*: Photo Russ Kleinman and Karen Blisard, Carlsbad Caverns Nationak Park, Grammer Seeps, May 14, 2014) (https://wnmu.edu/academic/nspages/gilaflora/f_hygrometrica17.jpg).

- It is demarcated from the theca proper by a constriction bearing a diaphragm (rim) composed of radially elongate cells.
- The rim joins the peristome to the epidermis of the wall.
- Above the rim is annulus with few of the cells distended. At the time of dehiscence these cells get disintegrated.
- Below the edge of the diaphragm is peristome **(Fig. 3.22)** which is made up of two rows of curved, narrow, triangular teeth, with sixteen in each row.
- Peristome are made up of cells joined together, and are arthrodontous type.
- The sixteen peristome teeth (exostome) are red and ornamented. The ornamentation is in the form of thick, transverse bands **(Fig. 3.23, Plate VI.11)**.
- The sixteen teeth converge at the tip and remain joined to a central disc **(Fig. 3.22).** Each tooth is a two ply structure and has two layers where the outer layer responds to changes in moisture content. This layer lengthens when wet and shortens when dry. These peristome have ability to bend inwards and outwards to regulate the spore dispersal **(Plate VI.12).**
- The inner row of teeth does not exhibit such hygroscopic changes and do not incurve (or outcurve).Teeth of inner peristome (endostome) are colourless, short, and delicate. They tightly fit into the spaces between the outer peristome.

Capsule

The middle fertile theca, from centre to outside, consists of a sterile columella,

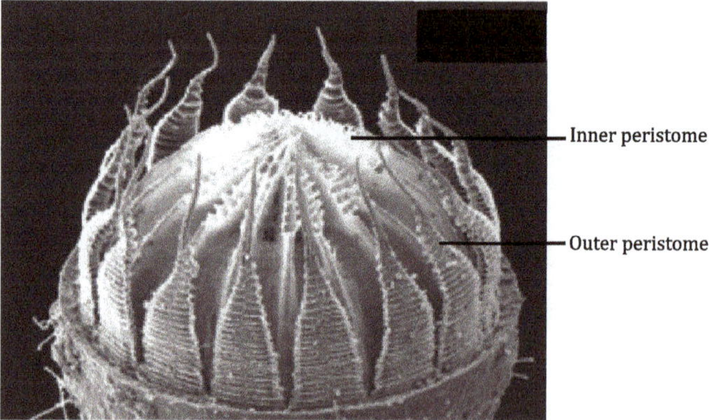

Figure 3.22 *Funaria* Photograph showing two rings of peristome. Spores are released through the spaces between the teeth. The inner peristome forms a dome over the mouth and keeps the mouth closed.

Figure 3.23 *Funaria* Peristome is arthrodontous where the outer teeth are articulated. Under dry conditions, the outer peristome which is hygroscopic, opens the capsule for shedding spores that are carried by wind; under moist conditions, the peristome keeps the capsule closed and spores are not shed. It however, acts like a sieve. (Source: www.google.com).

surrounded by a barrel-shaped spore sac, a cylindrical air space with trabeculae, hypodermis and epidermis.

- The capsule wall is multi-layered with outermost well defined epidermis followed by a two or three layered hyaline celled, hypodermis.
- Hypodermis is followed by innermost wall layer composed of loosely arranged cells rich in chloroplasts. It is photosynthetic in nature.
- In the centre is a sterile columella surrounded by barrel shaped spore sac (U-shaped) broken at the base.
- The spore sac has an outer wall of three or four layers of cells which lack chloroplast. The inner wall is single layered.
- The spore sac encloses haploid spores at maturity; no other sterile tissue is present.
- The spore sac is joined to the capsule wall by trabeculae which traverse the air space outside the spore sac. The trabeculae are formed from cells rich in chloroplasts and remain stretched at maturity **(Fig. 3.20).**

Apophysis

- This is a region where seta gives way to the capsule and is the base of the capsule.
- It is constructed with centre most tissue as the conducting strand, surrounded by a broad zone of spongy green tissue enclosing prominent air spaces.

- The outermost tissue is the epidermis with true stomata.
- The highlight of apophysis is its ability to act as an assimilating system with its chlorophyllous cells, intercellular spaces and functional stomata. The ability to photosynthesize makes the sporophyte of *Funaria* quite independent of the gametophyte from physiological point of view.

Dehiscence of Capsule and Dispersal of Spores

When the capsule dries up due to outside atmospheric conditions, operculum is thrown off to expose peristome consisting of two overlapping rings of peristome teeth. Each ring of peristome possesses sixteen teeth. The teeth of outer ring (exostome) are conspicuous, red with thick transverse bands whereas the inner rings (endostome) are comparatively small, colourless and soft. The dispersal of spores is due to hygroscopic movements (*viz.* movement due to moisture content of atmosphere) of exostome of peristomial teeth. The inner ring of peristome teeth does not show hygroscopic movements and functions as a sieve allowing only a few spores out of the capsule at a time. The outer peristome forms a dome not allowing spores to be set free under moist conditions **(Plate VI.12, VIII.5).** Spore dispersal in this highly advanced moss is not only with a few spores at a time but it occurs over a longer period and to a greater distance. This phenomenon is facilitated by hygroscopic seta which twists and untwists, jerking the spores from the spore sac and is accompanied by the swaying movement of seta with breeze that carries the capsule with it. Thus, spores produced are dispersed far and wide leading to less competition amongst the spores and their better survival (flow chart).

Protonema

- The spores germinate to form a branched, multicellular and filamentous structure, the primary protonema which grows through activity of an apical cell **(Plate VI.13)**.
- The protonema grows in two directions, some filaments grow prostrate and a little upright while the others grow down into the soil. The former are chlorophyllous and form the chloronema while the latter are rhizoidal.
- The chloronemal branches are thick with hyaline cell walls, the cross walls are at right angles to the lateral walls. Due to the presence of chloroplasts, these branches are photosynthetic.
- The rhizoidal branches appear thin walled, are brown in colour and have oblique cross walls. The cells abound in leucoplasts. These branches are meant to anchor the plant.
- According to Sironwal (1947), there are two clear cut phases in protonema development, the first stage is chloronemal, succeeded by the caulonema.
- The chloronema shows sparse and irregular branching, is positively phototrophic, does not form buds and produces a thermolabile growth promoting substance.

Capsule Dehiscence and Spore Dispersal

Mature capsule begins to dry up

↓

Columella and thin walled tissues lose water and shrivel

↓

Pressure builds up and tears up the spore sac

↓

Operculum removed assisted by annulus when moisture is available (hygroscopic)

↓

Annulus rolls back

↓

Operculum is thrown off

↓

Spore dispersal by peristomial movements

↓

Since outer peristome is a two ply structure

↓

When wet, the outer wall layer lengthens and the dome is created over spore sac and spores cannot be released

↓

When dry conditions prevail, outer wall shrinks, outcurving in the process

↓

This outcurving occurs with a jerk because the inner peristome fits into the outer and spores are exposed.

↓

The inner peristome does not show any movements but acts as a sieve

↓

Allows gradual dispersal of spores

↓

Seta assists through hygroscopic movement → Few spores at a time. ← Seta assists when it sways with wind

- After a few days, the chloronema gives rise to caulonema with profuse branching, oblique cross walls, walls brownish, negatively phototrophic, bears buds and forms a thermostable growth inhibiting substance.
- Thus, protonema has a distinct heterotrichous habit **(Fig. 3.24A).**
- Buds arise behind the cross walls and organize an apical cell which divides giving rise to a leafy gametophore **(Fig. 3.24B).**

**The sporophyte in bryophytes (Funaria) has begun to be independent of the gametophyte. Due to the presence of chloroplasts and intercellular spaces in the capsule wall and stomata in the apophysis, the sporophyte has acquired the potential to make its own food.*

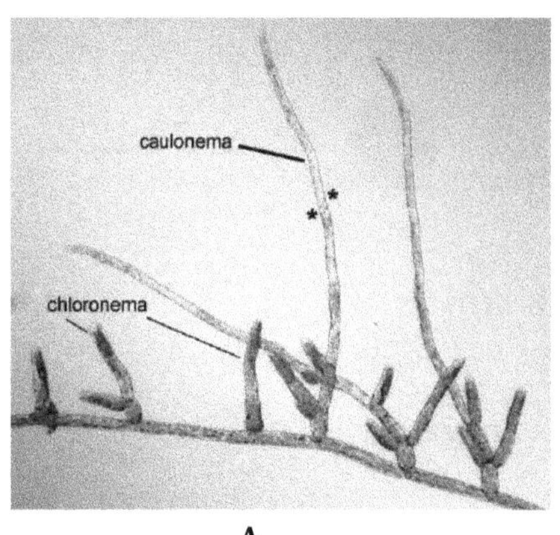

A

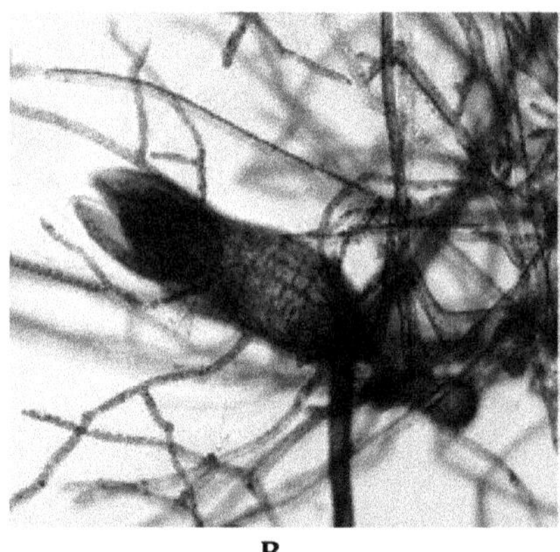

B

Figure 3.24*A&B Funaria* W.M. Protonema. *A* - Shows the prostrate system and the erect system. Also seen is the caulonemal system which has distinctive oblique cross walls. *B* - Shows leafy gametophore.

Check Your Knowledge

1. With the help of suitable diagram describe the structure of sporophyte of *Funaria*.
2. Spore release in *Funaria* is a gradual process. Justify.
3. Sporophyte of *Funaria* is partially independent. Comment.
4. Differentiate between the nematodontous and arthrodontous peristome.
5. Write short notes on:

   a) Moss protonema

   b) Archegonial 'head'

   c) Peristome teeth

Life cycle is depicted in **Figure 3.25.**

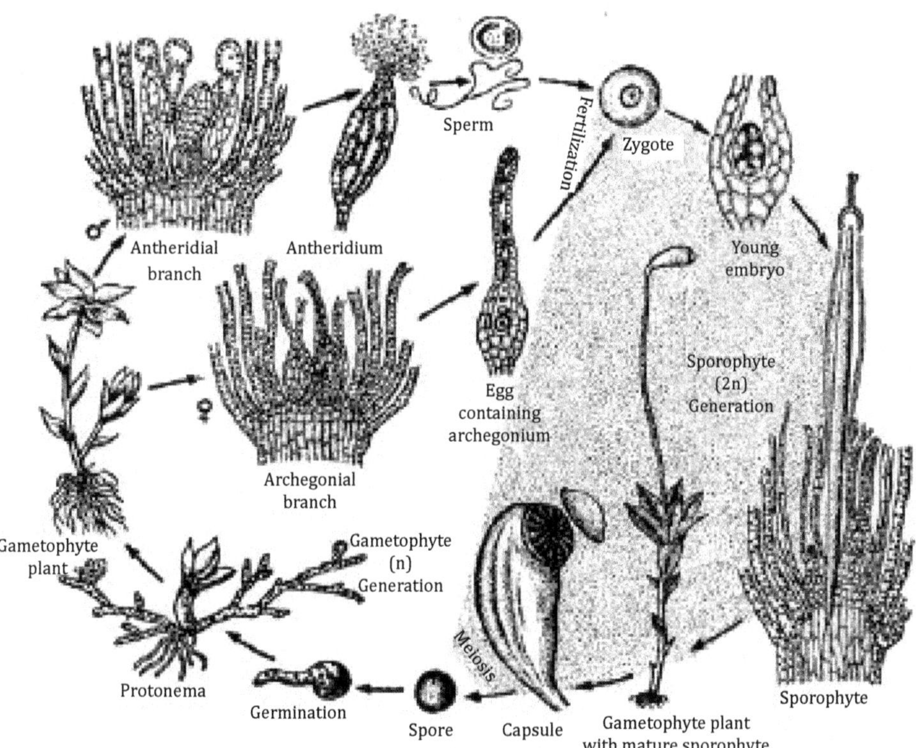

Figure 3.25 *Funaria* life cycle.

Chapter 4

Phylum– Anthocerotophyta – *The Hornworts*

The group is both unique and interesting in many features and researches reveal it to be distantly related from other bryophytes. Genetic evidences present hornworts as more like ferns (Groth-Malonek, 2005). Yet, chemical evidence as lack of isoprene emission, places them close to the liverworts (Hanson *et al.,* 1999) (mosses and ferns possess isoprene emission). However, such characters may prove to be retained or lost adaptively and contribute little to phylum level relationships. Hornworts represent a key group in the understanding of evolution of plants because the members are hypothesized to be sister to the tracheophytes (Qiu *et al.,* 2006). There are about 200-250 hornwort species in the world, which is fewer than either liverwort or moss species numbers (Villarreal *et al.,* 2010; Garcia *et al.,* 2012; Villarreal *et al.,* 2012). The hornworts are divided into two classes (Stotler & Crandall-Stotler, 2005), a concept supported by molecular data (Frey & Stech, 2005). Anthocerotopsida is the largest and best known of these, with two orders and three families. The second class Leiosporocerotopsida is monotypic, with one order, one family, and one genus. Hornworts have been frequently referred to as a synthetic assemblage with plants bearing resemblance to algae, liverworts, mosses and pteridophytes like *Rhynia*. They also have features not met with in any other group and have been enumerated in several literature:

- Hornworts differ from Marchantiophyta in having typically only one chloroplast per cell in the thallus, lacking oil bodies, and possessing a pyrenoid (a proteinaceous body serving as a nucleus for starch storage and common in green algae) in many taxa, which permits plants to assimilate CO_2 due to high concentration of Rubisco.

- The archegonia are produced with a distinct neck but diffused venter which is a character shared with pteridophytes.
- The sporophyte shows similarities to *Sphagnum* in lacking a sporophyte stalk (seta) and in having a columella that is not present in liverworts (but is seen in most mosses). The capsule wall also has functional stomata surrounded by two kidney-shaped guard cells, a character shared with mosses.
- Instead of elaters as in liverworts, the hornworts have pseudoelaters (arising from division of a pseudoelater mother cell) and outnumbering spores. They are formed of one, two, or four cells, usually with no spiral thickenings except *Megaceros* and *Dendroceros* (Renzaglia, 1978). The pseudoelaters probably provide nutrition, at least initially, but at maturity they show twisting movements, contributing to dehiscence and spore dispersal (Renzaglia, 1978). Continuous meiosis near the meristematic zone is also a unique feature which enables the spore production through out the life span of the sporophyte.
- Dispersal occurs when the capsule splits into two valves from the top down, and consistent with its development, this peeling back of the capsule occurs slowly over time, retaining the lower spores while dispersing the upper ones. The valves twist in response to moisture changes, aiding in dispersal. The spores mature progressively from top to bottom as the capsule splits and sporogenesis continues over a long period of time.
- Spores in Anthocerotophyta germinate to form a short protonema that does not remain thread-like, but gets more three-dimensional, resembling a tuber. This is not met with in other bryophytes.
- The mature gametophyte thallus resembles that of a club moss (Lycopodiophyta) in that the antheridia may occur in groups within a chamber. The archegonia are likewise embedded within the thallus, again like those of the club mosses.

Class Anthocerotopsida Jacnz. ex Stotl. & Crand. Stotl.

- It is an isolated evolutionary line and appears distinct from other bryophytes to the extent that a separate Phylum Anthocerotophyta has been constituted (Stotler & Crandall-Stotter, 2006).
- The group got its name as hornwort from the similarity of the sporophyte shape to the 'horn'.
- The plants of this group are dorsiventral, lobed thalli attached to the substratum by smooth, unicellular rhizoids.
- The thallus consists of only one type of tissue and each cell has a single large disc-shaped chloroplast with a pyrenoid.

- The thallus is frequented by mucilage filled cavities which are inhabited by symbiotically associated cyanobacteria.
- The thallus has slime pores on the ventral surface.
- Sex organs arise from the hypodermal cells and appear embedded in the thallus at maturity.
- The antheridia arise in numbers inside a single chamber and each antheridium is a discrete structure.
- In archegonium, venter is indistinct.
- The zygote undergoes a longitudinal division unlike other bryophytes where the division is transverse.
- Sporophyte lacks a distinct seta and due to the presence of basal intercalary meristem, it undergoes indeterminate growth. It remains attached to the gametophyte by a distinct foot.
- The spores present towards the tapering apex are mature and those towards the meristem are young.
- There is continued growth and production of spores from the archesporium. This is attributed to the basal meristem.
- Various stages of spore development (archesporium → sporogenous tissue → spore mother cells → spore tetrad → spores) are seen through the length of the sporophyte.
- The young sporophyte is enclosed inside calyptra or more appropriately involucre which gets ruptured at maturity.
- The dehiscence occurs by two valves that proceed apex downwards.
- The capsule jacket is multistratose and possesses functional stomata.
- The sterile columella is present in the centre of the capsule and is overarched by sporogenous tissue.
- The capsule encloses both spores and pseudoelaters.
- Spore dispersal occurs over a long period of growth.

Subclass Anthocerotidae Rosenv. corr. Prosk. emend Duff *et al.*

Order Anthocerotales Limpr. in Cohn.

Within Anthocerotopsida, two orders - Anthocerotales and Notothyladales have been recognized. Anthocerotales include a single family Anthocerotaceae with *Anthoceros, Folioceros* and *Sphaerosporoceros.* The order includes highly ornamented and darkly pigmented spores, formation of schizogenous mucilage cavities in dorsal half of the thallus and tiered antheridia that occur in large groups and open apically.

Family Anthocerotaceae Dumort. corr. Trevis. emend Hässel

Anthoceros

Habitat and Distribution

Anthoceros is the largest genus of hornworts, with ca. 83 species (Villarreal *et al.,* 2010) and about 25 species that are distributed in India. With a global distribution, the centres of diversity in the genus are in the Neotropics and tropical Africa and Asia.

The word 'Anthoceros' means 'flower horn', and refers to the characteristic horn-shaped sporophytes that all hornworts produce. The genus *Anthoceros* was first established by Micheli as early as 1729 and later was adopted by Linnaeus (1753). Three commonly found species of *Anthoceros viz., A. himalayensis, A. erectus* and *A. chambensis* are found growing luxuriantly in the Western Himalayan region at an altitude of 5000-8000 feet. These species are also found growing in Mussoorie, Kulu, Manali, Kumaon, Chamba valley, Punjab, Madras and in plains of South India.

All the species of *Anthoceros* grow on moist clayey soil and on wet rocks in very moist, shady places, usually in dense patches. Portions of moist slopes along the hill side roads, ditches, along claybanks of streams and moist hollows among rocks are the usual habitats where plants are commonly found. A few species of *Anthoceros* occur on decaying wood.

Gametophyte

The haploid spore germinates to form a filamentous or more commonly, tuberous structure which grows into an orbicular or sub-orbicular gametophyte with internally a uniform tissue. The thallus appears dark green because of its association with *Nostoc*, the cyanobacterium. Antheridia and archegonia are produced in chambers on dorsal surface of the thallus. Fertilization between egg and antherozoid results in zygote formation, thus begins the sporophytic generation. A highly differentiated sporophyte is produced from which spores are dispersed over a long period of time.

Morphology

- The thallus is dark green orbicular to sub-orbicular and may be smooth or may possess lamellae on dorsal surface. In the reproductive phase, thallus shows 'horn-like' sporophytes **(Fig. 4.1; Plate VII.1).**

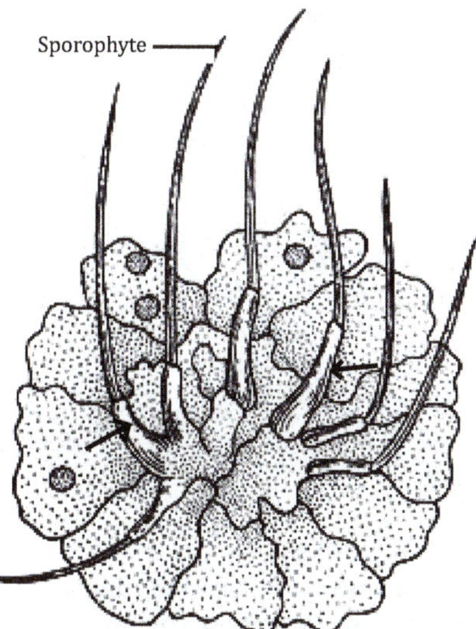

Figure 4.1 *Anthoceros* Diagrammatic sketch of thallus with sporophytes that stand out like 'horns'. Also seen is the pouch-like involucre (arrows) around the sporophyte.

- The ventral surface of the thallus appears dark bluish green because of the association with cyanobacteria **(Plate VII.2).**
- Branching is not distinctly dichotomous and thallus remains attached to the substratum by simple rhizoids.

Anatomy

- The epidermal layer leads to several celled thick parenchymatous region. Each cell bears a chloroplast associated with a pyrenoid **(Plate VII.3)**. This being an algal character, is unparalleled in bryophytes and makes *Anthoceros* and other members of Anthocerophyta share the character with algae like *Coleochete.* This is of evolutionary significance and is one of the arguments in support of algal ancestory of bryophytes.
- Towards the ventral surface there are mucilage cavities, which open outside by slime pores **(Fig. 4.2).**
- These cavities are inhabited by *Nostoc* which helps plant fix nitrogen.
- As the thallus matures, the mucilage in these cavities gets dried up **(Plate VII.4)** and the cavities are then filled with air.

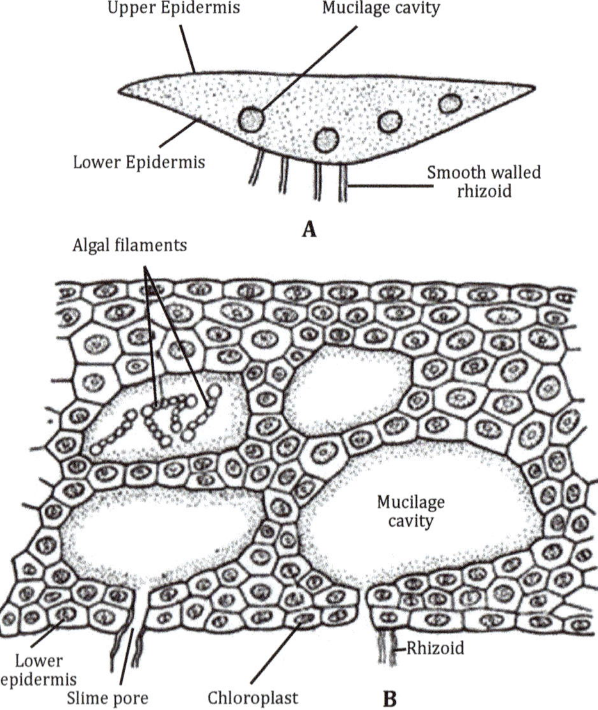

Figure 4.2*A&B Anthoceros* V.S.Thallus. *A* - Outline diagram to show rhizoids, mucilage cavities and epidermis intercepted by slime pores on the ventral surface. *B* - Cellular details to show tissues organized into upper and lower epidermis and inner parenchymatous cells with a single chloroplast per cell. The mucilage cavities at maturity may be filled with slime and are referred to as slime pores, while the young cavities are inhabited by *Nostoc* colonies.

Reproduction

Vegetative Reproduction

By Death and Decay of the Older Portion of the Thallus or Fragmentation

The older portions of the thallus rot or disintegrate due to ageing or dryness. As decay reaches up to dichotomy, the lobes of the thallus get separated. Detached lobes then develop into independent plants by apical growth. However, this method is not very common in *Anthoceros* as it is in liverworts.

By Tubers

Under unfavorable conditions and prolonged drought, the marginal tissues of the thallus get thickened to form perennating structures, the tubers **(Fig. 4.3).** Their position varies in different species and may develop behind the growing points (*A. laevis*) or along the margins of the thallus (*A. hallii, A. pearsoni*). In *A.himalayensis,* the tubers are stalked and develop along the margins on the ventral surface of the thallus **(Plate VII.5).** The tubers have outer two to three layers of corky hyaline cells which enclose cells containing oil globules, starch grains and aleurone granules. On arrival of favourable conditions tubers produce new thalli.

By Gemmae

In *A. glandulosus, A. propaguliferus, A. formosae,* many multicellular stalked gemmae develop along the margins of the dorsal surface of the thallus. When detached from the parent thallus, each gemma develops into a new plant.

By Persistent Growing Apices

Due to prolonged dry summer or towards the end of the growing season, the whole thallus in some species of *Anthoceros* (*A. pearsoni, A. fusiformis*) dries up and dies except the growing point. Later, this apical portion grows deep into the soil and becomes thick under unfavorable conditions. On return of favourable conditions, it develops into new thallus. It is more commonly followed to tide over unfavourable conditions.

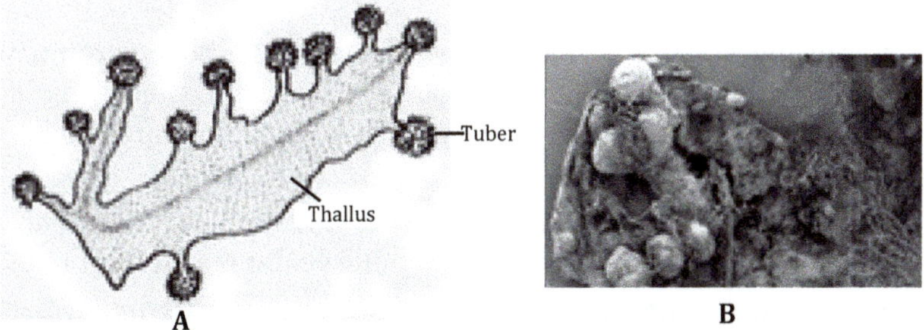

Figure 4.3*A&B Anthoceros. A* - Thallus showing tubers along the margins. The tubers are thick-walled perennating structures meant for vegetative propagation. *B* - Scanning photograph of tubers in a hornwort (source: Hornwort Heaven The authors, Silvia Pressel, Jeff Duckett and Juan Carlos Villarreal, recount their expedition into the Himalayan foothills of northern India: a thalloid Shangri-La).

By Apospory

Unspecialized cells of any part of the sporophyte (basal meristematic zone, sub epidermal and sporogenous region of the capsule) form the gametophytic thallus without forming spores. This phenomenon is called apospory (Lang, 1901, Schwarzenbach, 1926). The thalli are diploid but normal in appearance *e.g., A. laevis.*

Sexual Reproduction

Some species like *Anthoceros longii, A. gollani, A. fusiformis, A. punctatus, A. crispulus* and *A. himalayensis* are monoecious while others like *A. erectus, A. chambensis, A. hallii, A. pearsoni* and *A. laevis* are dioecious. The monoecious species are protandrous *i.e.,* antheridia mature before archegonia.

Antheridia

- Antheridia are produced on the dorsal surface inside antheridial chamber **(Plate VII.6).** Occasionally, mucilage is seen around the chamber.
- The antheridial chamber is quite distinct with a two layered roof and is filled with mucilage.
- Generally two to four antheridia are formed in a chamber with one, primary antheridium while rest arising from the stalk of older antheridia or the basal portion of the chamber, are the secondary antheridia. As a result, all antheridia in a chamber are of different age **(Plate VII.7).**
- The mature antheridium is club-shaped with a short but stout stalk. The cells in the uppermost region are triangular with a narrow opening towards the apex. Each cell of the jacket possesses plastids.
- Single layered jacket encloses antherozoids with a linear body.

Archegonia

- The female sex organs like male are produced close to the growing points of the thallus.
- They are visible by the mound of mucilage which accumulates around them **(Plate VII.8).**
- The most distinguishing feature that separates hornwort members from other bryophytes, is that only the neck region is formed by the archegonial initial. The venter region does not arise from the initial and is confluent with the thallus.
- To provide protection to the egg, the surrounding vegetative cells of the thallus overgrow. Absence of a definite venter is a pteridophyte character.
- Thus, the archegonia are not very discrete in construction.
- A mature archegonium consists of two to four cover cells, an axial row of four to six neck canal cells, a ventral canal cell and an egg.

- Fertilization occurs as in other groups of bryophytes. The zygote first secretes a thick wall, increases in size and divides to form three tiers of cells. The first division of zygote is longitudinal to the axis of the archegonium.

Sporophyte

- It is a tapered cylinder attached to the gametophytic tissues below by a bulging parenchymatous foot **(Plate VII.9).**
- The foot which remains embedded in the gametophyte may produce short haustorial projections for better anchorage and absorption. When young, involucre around the sporophyte is intact **(Plate VII.10)**.
- Foot gives way to basal meristem (in place of seta) which continues to form two embryonic tissues, the amphithecium and the endothecium from which other tissues arise.
- The entire endothecium in the capsule region gives rise to columella which in mature sporophytes is made up of 16 vertical rows of cells and is largely associated with mechanical function **(Plate VII.11)**.
- The amphithecium gives rise to the rest of the capsule body - the multistratose wall, pseudoelaters and spores **(Plate VII.12).**
- The wall has outer epidermis with functional stomata and leads to inner zone of chlorenchymatous cells.

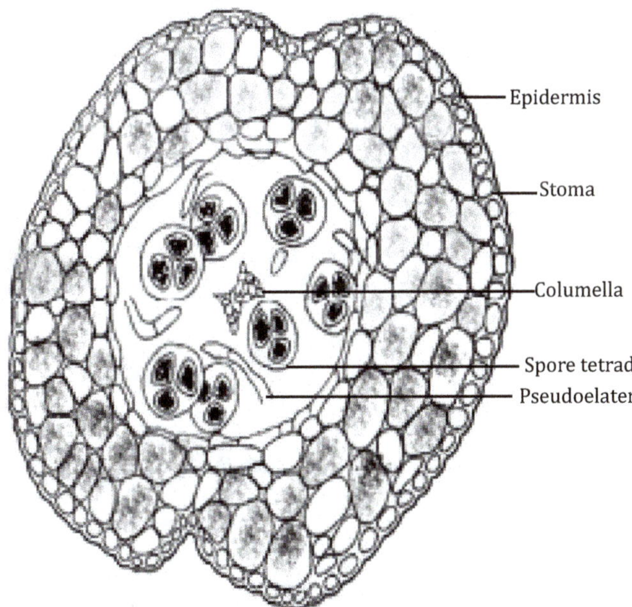

Figure 4.4 *Anthoceros* T.S. Sporophyte - Diagrammatic sketch to show multistratose capsule wall comprising epidermis with stomata followed towards inside by chlorophyllous tissue. Also seen is 16-celled columella surrounded by spore sac bearing spore tetrads and pseudoelaters.

- Pseudoelaters arise from pseudoelater mother cell. They are sterile, branched structures, without spiral thickenings but show hygroscopic movements. When young, they are photosynthetic, at maturity they help in spore dispersal **(Fig. 4.4; Plate VII.13).**

Capsule Dehiscence

- The young sporophyte remains enclosed by involucre of gametophytic origin. The base of the foot remains embedded in the gametophyte where transfer cells have been reported **(Fig. 4.5A).**
- Towards the tip mature stages of spore tetrads and spores are seen **(Fig. 4.5B).**
- As the tip matures, the capsule undergoes change in colour, gradually

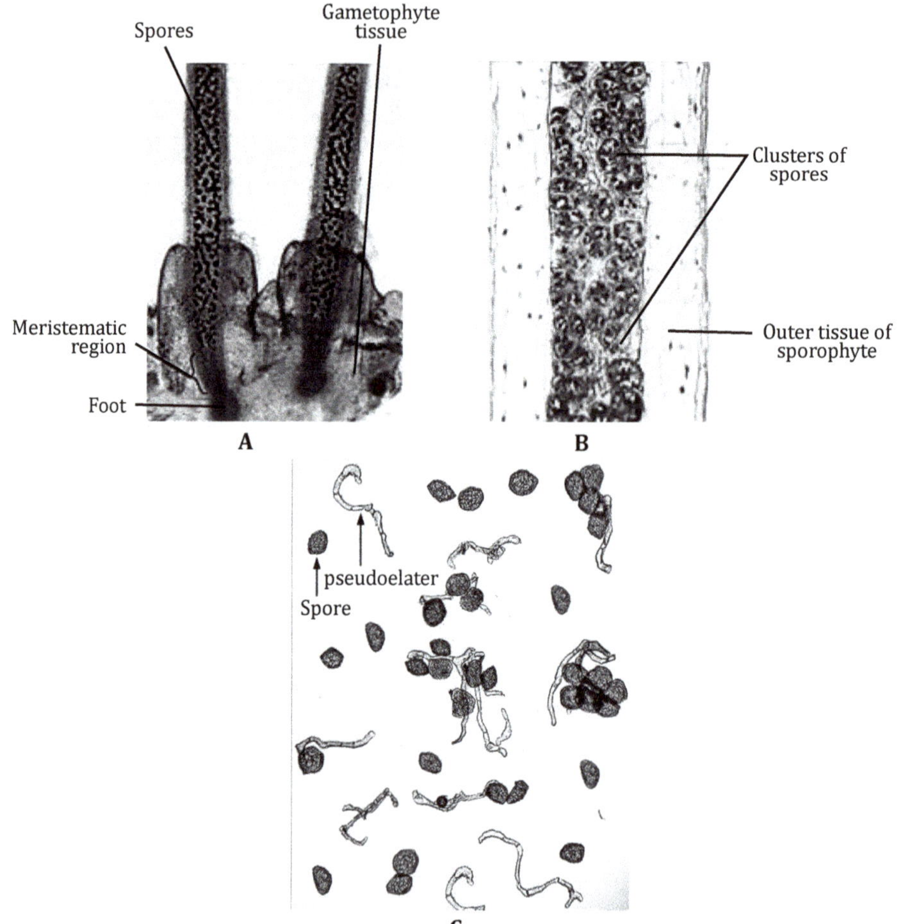

Figure 4.5*A-C* *Anthoceros* L.S. Sporophyte. *A* - Young sporophyte with foot embedded in the gametophyte. Also seen is the involucre. *B* - The mature stages in spore formation are seen towards the tip. *C* - The spores remain entangled with pseudoelaters that assist spore dispersal.

shrivels and splits forming two valves which are hygroscopic. The combined movements of these valves and pseudoelaters help disperse the spores **(Fig. 4.5C, Plate VIII.8A&B).**

- Due to the meristematic region, spore formation is a continuous process *i.e.,* while the tip releases spores, the base forms archesporium which through sporogenesis produces spores. Spore dispersal therefore, continues for a longer period **(Fig. 4.6A-D).**

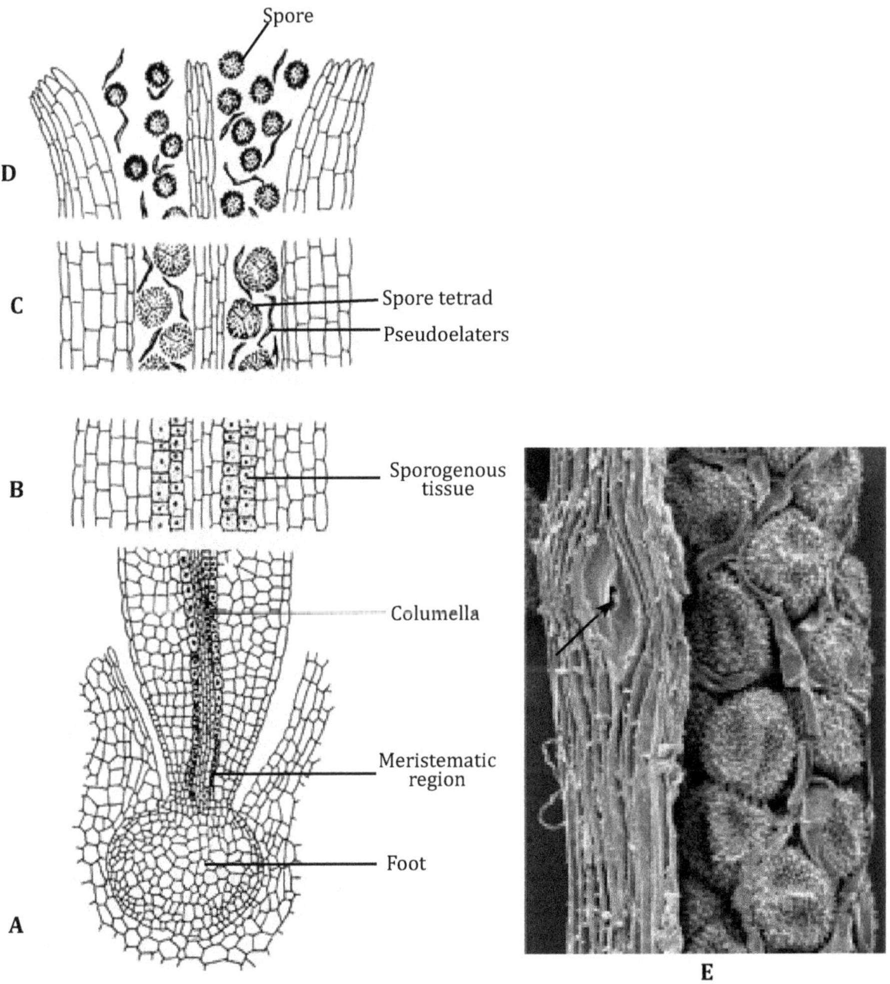

Figure 4.6*A-E Anthoceros* L.S. Sporophyte to show various stages in spore formation. *A*- The globose foot remains inside the involucre and is deeply embedded in gametophyte. The intercalary meristem is responsible for continuous spore production. *B* - Shows sporogenous tissue. *C* - Capsule at the spore tetrad stage. *D* - Spores and pseudoelaters. *E* - The capsule/sporophyte wall has functional stomata (arrow).

- The sporophyte in *Anthoceros* has stomata in the wall **(Fig 4.6E),** chlorophylous tissue and foot deeply embedded in the gametophyte where transfer cells are reported. Hence, the hornwort sporophyte is quite an independent structure.

Life cycle is depicted in **Figure 4.7.**

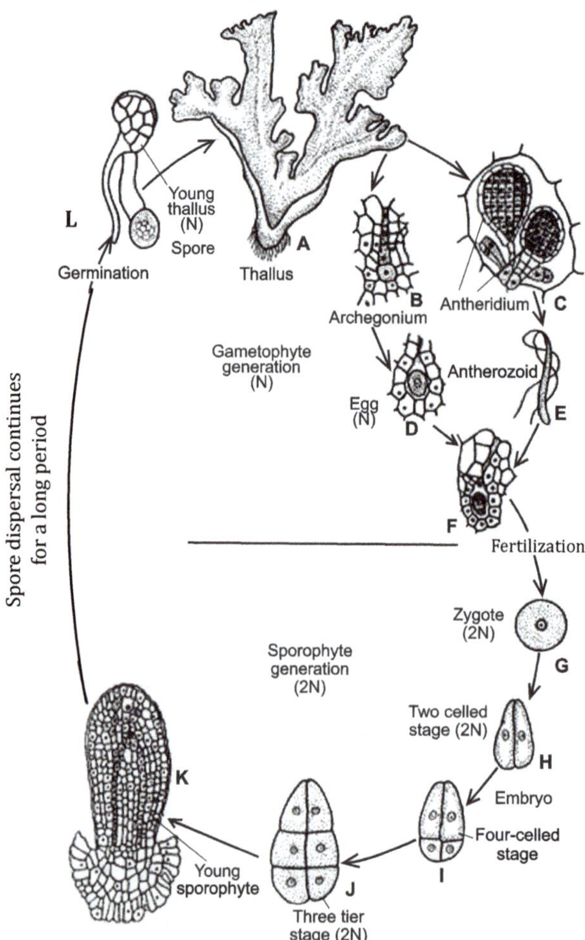

Figure 4.7 *Anthoceros* Life cycle showing distinct phases. *A* - Thallus, *B-F* - Male and female sex organs representing the gametophytic phase. *G-K* - The diploid stages. *M-O* - Formation of gametophyte.

Evolutionary Trends

Hornworts form an important group when progressive sterility of the sporophyte amongst bryophytes is considered. In liverworts, the entire endothecium gives rise to spores as in *Riccia* (only in some cases sterility is seen *e.g., R. crystallina* where sterile nurse cells co-exist with spores), and in *Marchantia, Pellia* and *Porella,* it forms spores and elaters (in *Pellia,* also an elaterophore is formed) adding to some sterility in the form of elaters. However, in *Anthoceros*, sterility in the capsule increases with entire endothecium devoted to formation of columella, a sterile tissue. The extent of sterility increases when amphithecium gives rise to multistratose wall of the sporophyte, and pseudoelaters, besides forming spores. This indicates that there is much more sterility in sporophyte of hornworts when compared to that in liverworts.

The group is also significant as it shares similarities with algae on one hand and pteridophytes on the other. It is perhaps these synthetic characters which relate it to ancestoral stock.

It is also an example of a highly differentiated sporophyte which has evolved towards independence. If the functional independence for nourishment is considered, sporophyte of *Anthoceros* has all the structures required for carbon assimilation. The epidermis with stomata that leads to chlorophyllose tissue, with intercellular spaces is an indication that the sporophyte can synthesize its own food. Further, columella which is occasionally implicated as mechanical tissue has been associated with water conduction in *A. fusiformis* (Campbell, 1924). However, sporophyte is still not an independent structure. This is because the sporophyte is not able to anchor itself in soil but remains embedded in the gametophyte like all other bryophytes. If the foot could become rooted in the soil, the sporophyte would be an independent structure.

Campbell (1924) reported a situation close to this assumption in *A. fusiformis* from California. Certain sporophytes of this species were about six inches tall and had a diameter almost double than the normal ones, columella was very well developed, chlorophyllose tissue was also distinct and foot in some cases reached the substratum after disorganizing the gametophytic tissue. In such cases, the sporophyte resembled those of Rhyniaceae, a pteridophytic group of fossil forms with *Rhynia* as a very close relative. Both are rootless, leafless, dichotomously branched, and bear terminal sporangia. This is progressive evolution in which from simple forms like *Rhynia*, other complex forms evolved. Other school of thought which believes in retrogressive evolution keeps simple members of Psilophytales (such as *Rhynia*) at the top and propose that the first pteridophytes were complex and must have undergone simplification to form bryophytes. However, it has been strongly proposed that *Anthoceros* is a connecting link between bryophytes and pteridophytes because latter are of anthocerotean origin **(Box 4)**.

Innovations in Hornworts

- Thallus is thickened with no internal differentiation except for extensive mucilage canals, colonized by *Nostoc* symbionts through epidermal clefts.
- The cells of the thallus possess chloroplasts with pyrenoid. The thylakoid system is unique in its arrangement (Renzaglia *et al.*, 2009).
- The consistent association with nitrogen fixing cyanobacteria and pyrenoid-based carbon dioxide concentrating mechanism is also of advantage to the plant.

Box 4

The well-preserved fossil remains of Rhyniopsida in rocks of the Silurian and the Devonian periods throw light on the "Dawn land plants" a land mark discovery of the 20th century. It is beyond doubt that the first land plant is *Cooksonia* of the late Silurian period. This early plant was "root less and leafless", and was of low height with stem dichotomies terminating in sporangia. At the end of the Silurian there began a rapid burst of evolution in plants. Within just 25 million years during the Devonian period, recognised as being one of the two intervals of greatest expansion in plant evolution, land plants evolved complex vascular systems, leaves which specialised in photosynthesis, and roots and stems for support. Most significantly sporangia were developed and seed-bearing plants evolved. Although *Cooksonia* evolved earlier, many consider *Rhynia* to be the first true vascular terrestrial plant. Two species of *Rhynia*, namely *R.major* and *R. gwynne-vaughanii* were discovered with a variation only in their size, the latter species being a smaller version of the former, while all other characters were similar. A re-examination of the original material of *Rhynia gwynne-vaughanii* retrieved from the Rhynie chert in Aberdeenshire, Scotland showed that the plant axis of the species was gametophytic. Even earlier botanists Kidston and Lang (1917) could never trace the actual connection between axis of *Rhynia gwnne-vaughanii* with its sporangium, presence of which could be a sure indication of diploidy of the axis. Pant (1962) suggested that fossil plant resembled more a gametophyte than a sporophyte. Lemoigne (1968) demonstrated that the transection of the plant axis showed the presence of archegonia. On this basis it was deduced that *R. gwynne-vaughanii* is the gametophytic generation of *R. major*. The axis *of R.gwynn-evaughanii* had a strand of tracheids representing vascular strands as in the extant gametophyte of *Psilotum*. These observations provided documentation for the homologous theory of the origin of alternation of generations. This added credibility to the belief that the sporophytic and the gametophytic generations of primitive vascular plants were similar in morphology.

Rhynia is an ideal plant for understanding evolution in the vascular plants. In having the simplest vascular trace, *Rhynia* is the starting point for Zimmermann's Telome Theory of the evolution of vascular system. The study of *Rhynia* suggests that the ancestor vascular plants were just stems, and all other structures, modifications of stem.

Leclercq (1954) for the first time remarked that Dewonian Rhynie flora in general and *Rhynia* in particular was probably not a representative of the earliest land plants. This was because, *Asteroxylon* which co-existed with *Rhynia* was a more complex form. Therefore, the primitive status of *Rhynia* was questionable. However, it was a simple plant. It is possible that *Rhynia* after attaining its present structure did not evolve further but continued to exist in this form for million of years. This might force us to believe that it is truly primitive. Alternatively, if *Rhynia* is reduced form from more complex land plants then it is not a primitive plant. Thus, whether *Rhynia* should be considered primitive remains inconclusive.

More significant about *Rhynia* was its simple nature which was unquestionable. It lacked differentiation into stem and roots and showed uniform dichotomous branching. Its appendages could not be called leaves. The stele was a simple strand of tracheids. The sporangia were modified branch tips. However, it its simple construction draws our concept of a vascular plant. Therefore, irrespective of whether *Rhynia* is a reduced form from complex ancestors or is truly a primitive form, in its simple organization, it represents one of the 'stages' in the evolution of vascular plants.

- There is total lack of water conducting cells as seen in mosses.
- In contrast to liverworts and mosses, hornworts lack external appendages in the form of leaves and scales.
- Vegetative reproduction involves additional terminal tubers.
- The gametangia show clefts of mucilage which is also present near apical meristem.
- Antherozoids are linear in shape while the archegonia have venter confluent with gametophyte thallus as in pteridophytes.
- The first division of the zygote is longitudinal to the long axis of the archegonium and subsequent divisions lead to the three tired embryo. The lowest tier gives rise to the haustorium- like foot while the uppermost gives rise to the spore bearing capsule; both these tiers undergo limited number of divisions. The middle tier forms the meristematic region which continues to cut off embryonic cells for extended length of time.
- Capsule wall has stomata with cells similar to guard cell of higher plants.
- The placenta which is involved in providing nourishment to the sporophyte has gametophytic transfer cells and sporophytic haustorial cells. This is a big achievement as the land plants – the tracheophytes slowly evolved from this type with sporophyte phase as the dominant one.
- Seta is lacking and sporophyte grows by means of a persistent basal meristem. This affords a more extended reproductive season than in mosses and liverworts.
- Sporogenesis is non synchronous and is long lived.
- Spore colour is indicative of longevity where yellow and brown spores are long -lived (because of thick walls and stored oils), green spores being thin walled are short- lived and there are black spores.

Check Your Knowledge

1. Write short notes on:
   a) Sporophyte of *Anthoceros.*
   b) Evolutionary significance of *Anthoceros.*
   c) Advantage of basal meristem in the sporophyte.
2. Why are elaters of *Anthoceros* called pseudoelaters?
3. *Anthoceros* is a link between bryophytes and pteridophytes. Comment

Chapter 5

Bryophytes and Human Welfare

A general lack of commercial value, small size of unpalatable thalli, perhaps with questionable nutritive value and inconspicuous place in the ecosystem, are some of the considerations that have made bryophytes appear to be of practically no use to most people. However, literature cites that these plants have been used in past for their value in ecological studies, fuel industry, medicines and even aesthetics and much more. Both liverworts and mosses are often good indicators of environmental conditions and have been used as indicators of past climate. Mosses are often used to condition the soil and are preferred in organic farming. Coarse textured mosses increase water-storage capacity, whereas fine-textured mosses provide air spaces to the soil (Ishikawa, 1974). Mosses improve the nutrient condition by holding nutrients, especially those borne by dust and rainfall, and releasing them slowly over a much longer period of time than normal nutrient composition near the soil surface (Stewart, 1977; Rieley *et al.,* 1979). Bryophytes in aquaria provide oxygen, hiding places, and egg-laying substrates for fish. In Germany, *Sphagnum* is used to line hiking boots (Hedenäs, 1991), where it absorbs moisture and odor. Thus, these seemingly unimportant plants have a lot to offer to mankind.

Food

Bryophytes are not directly used as food because most of the species are not palatable. *Sphagnum* is used as a wretched food in barbarous communities. In China, peat moss is occasionally used as a famine food. It is also used as ingredients in the preparation of bread. *Sphagnum* has also been used in dietary requirement of animals. It is also supplemented with egg yolk for anuran tadpoles. The capsules of *Bryum* and *Polytrichum* are consumed by the Norwegian grouse chicks, whereas other capsules are consumed by Scottish red grouse. Even large prairie mammals are known to eat mosses. The mayfly, *Ephemerella* feeds on the green alga *Ulothrix* when it is available, but feeds on the ever-present moss when the alga is scarce or absent. Terrestrial insects likewise eat mosses; capsules seem to be a favoured part. Milled *Sphagnum* is an ideal binder for the iron and vitamins and is fed to baby pigs because of its ability to absorb and hold nutrients (Ando & Matsuo, 1984).

Industry

Moss is a soil conditioner and is used as plant- growing medium. *Sphagnum* is preferred all over the world for industry purposes. It holds up to 30 times its weight in water, making it invaluable for packing and shipment of plants and fresh vegetables and flowers. It is used in hydroponics gardening and for storage of roots and bulbs. Fishermen use *Sphagnum* to keep their worms alive. Powdery mass of *Sphagnum* is used in making peat wood and peat created are used as valuable construction material. Certain countries such as Sweden use peat in making wrapping paper and pasteboard (Ando, 1983).

Horticulture

Horticulture enjoys a long tradition involving bryophytes (Perin, 1962; Arzeni, 1963; Adderley, 1964) as soil additives, ground cover, dwarf plants, greenhouse crops, potted ornamental plants, and for seedling beds (Sjors, 1980). *Sphagnum* is used in making totem poles to support climbing plants and moss-filled wreaths, popular in southeastern U.S. Other decorative horticultural uses include making baskets and covering flower pots and containers for floral arrangements (Thomson, 1994). Mosses are frequently used as a nursery material for growing seedlings of plant species. They are used as orchid growing medium. Mosses are wrapped around the roots of seedlings and kept in the nurseries for hardening. Mosses provide appropriate moisture and congenial environment for seedling growth. Moss sticks are widely used for indoor climbing plants. Mosses are commonly used to make flowerpots and are an intrinsic part of the diversity and beauty of life. In Japan, growing mosses is a traditional part of horticulture. The bryophytes are being used in gardening and as ornamental material for cultivation in landscape trays. Mosses are also planted in "moss-gardens" particularly at Buddhist temples where mosses create an atmosphere of beauty, harmony and serenity reflecting the spirit of Buddhism.

Medicines

Moss species of *Philonotis, Bryum, Mnium,* and various matted hypnaceous forms were in ancient times used to alleviate the pain of burns. These mosses were crushed into a kind of paste and applied as a poultice. People in the Himalayas use burnt ash of mosses mixed with fat and honey that serves as an ointment for cuts, burns and wounds. This paste apparently provides both soothing and healing effects. Dried *Sphagnum* is sold in herb shops in China for treatment of haemorrhages (Ando, 1983). It is used in the treatment of boils and in medicinal baths. *Marchantia polymorpha* thalli in ancient times were used to cure ailments of the liver and jaundice. In China, it has been used to treat jaundice (hepatitis) and also as an external cure to reduce inflammation. It was also used in Europe to cure pulmonary tuberculosis. The rosette-forming *Riccia sps* is used as an external application, as, cure against ringworm. The extracted oil from *Polytrichum commune* is applied to the hair to strengthen and beautify it. In China, *Polytrichum commune* is boiled to make tea for treating cold (Pant, 1998).

Bryophytes have been used as crude drugs and 30 to 40 species have been recognized as having effective cures. *Rhodobryum giganteum* and *Rhodobryum roseum* are used for the treatment of cardio-vascular diseases and nervous prostration, *Polytrichum commune* to reduce inflamation as an anti-fever agent, while laxative, *Haplocladium microphyllum* for tonsillitis, bronchitis and cystitis, *Conocephalum conicum* and *Marchantia polymorpha* mixed with vegetable oils are used as ointments for boils, eczema, cuts, bites, wounds and burns. *Fissidens* is used as an antibacterial agent for symptoms such as a swollen throat. Many mosses such as *Hypnum cupressiforme* also have antifungal activity (Banerjee & Sen, 1979). Several antiviral, active humic acids are found in *Sphagnum.*

Trade

The use for intact moss is principally in the florist trade and for home decoration. In the remote areas, mosses are used as material for bedding, packing, plugging and stuffing owing to their soft elastic texture. The mosses are suitable for these purposes because they are not easily attacked by microorganisms and require no special treatment except drying out (Kumar *et al.,* 2000). Some mosses such as *Neckera menziesii, Dendroalsia abietina* and *Antitrichia california* are used as moisture retaining packing for vegetables. *Rhytidiadelphus triquetrus* has been used for packing fragile articles and silk clothes in China and Japan. Moist *Sphagnum* is used in packing live frogs, snakes, lizards, worms and insects for shipment. In India, mosses such as *Sphagnum* are frequently used for packing of apples. In Himalayas apples and plums are wrapped in mosses including *Brachythecium salebrosum, Cryptoleptodon flexuosus, Hypnum cupressiforme, Macrothamnium submacrocarpum, Neckera crenulata, Trachypodopsis crispatula* and *Thuidium tamariscellum.* Nurserymen in India use wet *Sphagnum* for sending or supplying live plants. It has been used in shipment of vegetables, trees, cacti, orchids, ferns, and other delicate plants. In Europe, draggers and scrapers are packed in mosses such as *Sphagnum, Plagiomnium undulatum* and *Hypnum. Pseudoscleropodium purum* and *Hylocomnium splendens* also provide packing and stuffing materials (Ando & Matsuo, 1984, Glime & Saxena, 1991).

Ecological Importance

Succession

Mosses along with lichens play an important role in the formation of soil and development of vegetation cover. Lichens are the pioneers to colonize barren and bare rocky surfaces where no other plants grow. The organic acid secreted by lichens gradually dissolves and disintegrates the rocks. The rock particles together with the dead and decaying older parts of the lichens contribute to soil fertility. Mosses make their appearance when this fertile soil gathers in the crevices of the rock surface. The progressive growth and death of the older parts of the mosses are added to the substratum. Dust and debris blown by the wind rapidly accumulate between the erect moss stems. The moss mat collects sufficient moisture and humus it contains, a suitable substratum for the growth of many rock loving herbaceous plants is formed. Thus, the moss stage follows the lichen stage and precedes the herbs, the latter in turn succeeded by shrubs and trees. Mosses play an important role in Bog succession from open water to climax forest. The mosses especially the peat mosses established on the banks of lakes and other shallow bodies of water extend inwards and grow over the surface of water with their stems interwined to form thick mats. These surface mats over bodies of water give the appearance of solid soil. Such areas are quacking bogs. The thick moss mat because of the moisture and humus forms a suitable substratum for the germination of seeds of various species of hydrophytic plants. The lakes and shallow water bodies become filled with partially decomposed old parts of mosses and other hydrophytic plants. Thus, the areas which were originally sterile sheets of water, become converted into soil supporting vegetation. The mosses and hydrophytic angiosperms eventually disappear and are replaced by forest growth of mesophytic type. The mosses thus, play a vital role in changing the landscape and in some cases even reclaim land.

Indicators of Environmental Health

Phenolic compounds isolated from *Plagiochila* and *Wiesnerella* are used as pesticides. Bryophytes have great capability to help in building the mineral rich rocks. It has been observed that some mosses, lichens and liverworts decrease or they show signs of injury in disturbed environmental conditions (air pollution). Most moss species disappear from such polluted areas except a few tolerant species *viz., Bryum, Ceratodon, Dicranoweisia, Stokensella* and *Tortula.* Relationship between the distribution of communities and Index of Atmospheric Purity (IAP) values can be calculated by periodic recording of injury, or any change in colour of the plants. The high concentration of Cadmium in some bryophytes shows a peculiar change in pigmentation and growth rate of these lower plants. In *Marchantia* and *Funaria*, Zinc concentration (>50 ppm) reduces spore germination. Some species of *Bryum, Dicranella* and *Polytrichum* are able to tolerate high levels of Zinc (5500 ppm), Cadmium (610 ppm) and Copper (2700 ppm) in their tissues.

Mosses show many morphological forms and colors due to which lawns and landscapes can be made to look beautiful and natural by planting these mosses. Bryophytes are indicators of a particular kind of site as species have a relationship with specific site. They are able to indicate the past existence of a forest or non-forest vegetation and can be used as indicator to reconstruct plantation of the particular area thereby acting as significant tools for restoration programme. They are also useful for biogeochemical panorama; the mosses such as *Milichhoferia, Merceya, Barbula* and *Scopelophila* are related with Cu-rich substrata. Some bryophytes are particularly grown on calcium-rich rocks and in such soils where waters flow over through calcicole rocks. Bryophytes help in precipitating calcium and form a mineral rich rock on one hand and filtering water on the other hand. Some taxa specifically grow on potassium – containing substrata. The patches of *Hyophila, Oxystegus* and *Zygodon* indicate iron hematite bearing substrates. Bryophytes are ideal pollution indicators because of their habitat diversity, structural simplicity, totipotency and rapid rate multiplication. Sensitive species show the change in plant morphology according to the environment quality. Mosses have great cation exchange properties because of their cell walls which have special chemical composition. Tolerant species serve as accumulator of the pollutants. They have a capability to absorb the pollutants in large quantity than absorbed by other higher plants of the same site thus, making them ideal organism for bioremediation programme. This absorbing capacity is due to the lack of cuticle, presence of single cell thick lamina and large surface by volume ratio while the other higher plants lack such characters (Glime & Saxena, 1991).

Miscellaneous

Bryophyte crusts endowed with nitrogen-fixing Cyanobacteria, can contribute considerable soil nitrogen, particularly to dry rangeland soils. Some of these Cyanobacteria are seen associated with *Anthoceros*. The moss *Bryum argenteum* is being used to monitor the thickness of the ozone layer over Antarctica (Hedenäs,1991). As the ozone layer decreases, increased exposure to UV-B radiation stimulates production of flavonoids in this species. In Japan, mosses are used to create a feeling of serenity in gardens. Instead of the mix of grass and flashes of color typical of western gardens, Japanese moss gardens have an uncluttered look of shades of green. Moss gardens are often associated with Buddhist temples, the most famous of which is Kyoto's Kokedera, literally translated as "moss temple."

Chapter 6

Evolutionary Trends

Origin of Bryophytes

Before molecular and phylogenetic studies on bryophytes became available, the morphological and structural characters were used to derive inter-relationships between various groups. Theories proposed were only speculative. It was generally agreed that bryophytes had forms that were more evolved than algae but less advance than pteridophytes. Therefore, if simple forms gave rise to complex ones, then evolution should have progressed from algae to bryophytes and from bryo- to pteridophytes. However, presence of both, simple and complex forms in liverworts and simple forms amongst mosses proved this line of evolution unconvincing to several workers. In light of such researches, two views were proposed to explain origin of bryophytes; Progressive evolution theory and Retrogressive evolution theory, while the third one, Transmigration theory remained insignificant.

Algal origin: Many workers (Campbell, 1940; Smith, 1955) believed that bryophytes originated from filamentous fresh water green algae. However, the fossil record did not support this hypothesis. Considering the similarity between filamentous green protonema with caulonema and chloronema of mosses, heterotrichous habit of algae as in Chaetophorales and presence of chloroplast with pyrenoid as in *Anthoceros*, the hypothesis could not be totally ignored. It was suggested that bryophytes and green algae have been derived from a common ancestor. There were two lines of evolution in green algae, the chlorophyceaen and the charophyceaen. The bryophytes and other archegoniates must have arisen from the charophycean line of evolution. In 1903, Lignier postulated that algal ancestors gave rise to a primitive terrestrial type, the Prohepatics from which arose bryophytes on one hand and vascular cryptogams on the other.

Pteridophytean origin: According to Takhtajan (1953) and Christensen (1957) simple vascular plants like Psilophytales have given rise to bryophytes by reduction. The theory places emphasis on the simplicity of the sporophyte that may have originated through considerable structural reduction from a branched sporophyte that bore terminal sporangia as in *Rhynia*. Support for this theory is generated from the sporophyte of *Anthoceros* which is comparable to *Rhynia*, plant with dichotomous leafless stalks which bore sporangia terminally.

Church (1919) gave his views in **'Thalassiophyta and the Subaerial Transmigration'**. According to him bryophytes arose from aquatic ancestors but which inhabited marine waters and not fresh water. These benthic forms developed sieve tube-like elements as seen in brown algae and were capable of photosynthesis. Such forms gave rise to land plants when the sea dried up and land emerged. There occurred *transition in situ*. This view was not accepted by most workers because of lack of evidence.

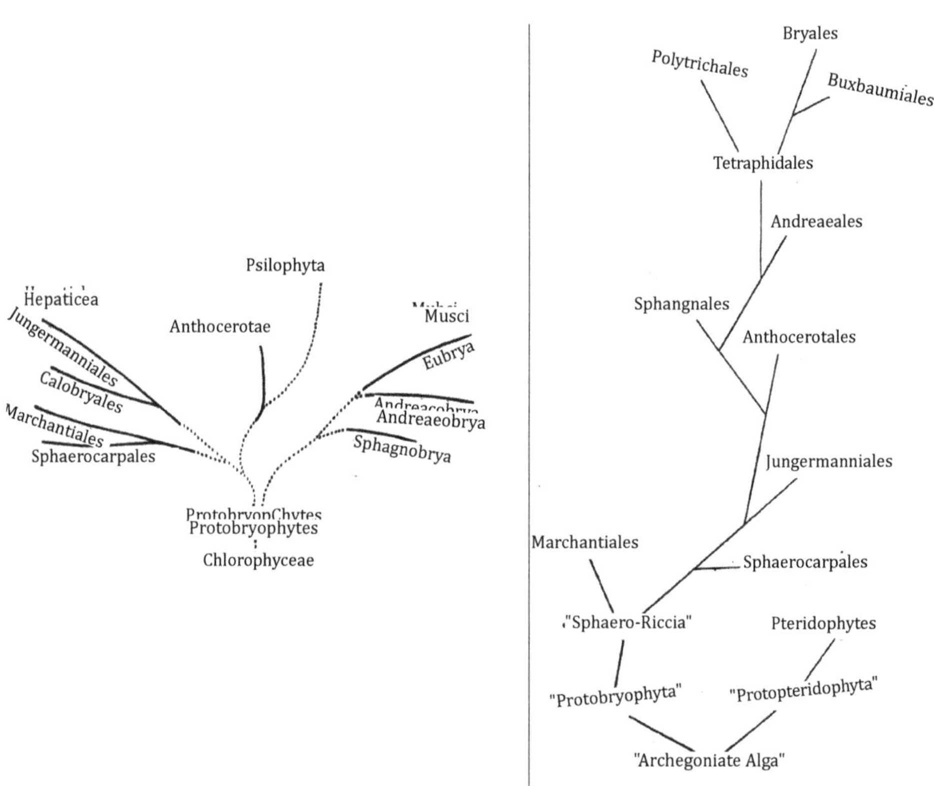

Figure 6.1*A&B A* - Diagram showing the suggested interrelationships among the Bryophytes. (*Source*:Smith 1955, Cryptogamic Botany). *B* - Phylogeny of Bryophytes (Cavers 1911).

Inter-relationships among Bryophytes: Early views

Three major evolutionary lines are recognizable among bryophytes – Liverworts, hornworts and mosses **(Fig. 6.1).** The consideration as to which form would resemble gametophyte and sporophyte of primitive ancestral form remained the main argument in proposing any hypothesis. One school of thought believed that a primitive gametophyte was a complex form and it slowly lost its complexities through stages and resulted in a more advanced though simple *Riccia* – like thallus. However, Smith from other school of thought believed that primitive bryophyte was simple like *Riccia* with both, gametophyte and sporophyte phases lacking any structural and functional complexities. According to Lotsy, protobryophyte could be a combination of simple gametophyte of *Sphaerocarpos* and simple sporophyte of Ricciaceae. Thus, at the bottom of evolution stood *Sphaero-riccia.* On the other hand, exist gametophytes of hornworts that resemble those of liverworts excepting the embedded sex organs in the former, but the indeterminate growth in hornworts is of an advanced type not seen in liverworts. The hornworts were considered to be a series which departed from primitive liverworts along the line that led to the Sphaerocarples and Marchantiales.

Progressive Sterilization of Sporogenous Tissue in Bryophytes!

The process of sterilization of sporogenous tissue through different groups is very significant in bryophytes. According to Bower (1890), more complex sporophytes have evolved from the simpler ones by the progressive sterilization of potentially sporogenous tissue. The evolution may be traced from the simplest sporophyte found in *Riccia* (liverwort) to the more complex sporophyte found in *Funaria* (moss). The discussion is based on two simple facts - First, the zygote undergoes a transverse division forming an inner hypobasal cell and an epibasal cell. These two cells participate in formation of the sporophyte (with foot, seta and capsule in advanced forms). Second, in the capsule region of the sporophyte, two fundamental tissues are formed – endothecium (inner layer) and amphithecium (outer layer). These layers are involved in columella, capsule wall and spore formation, the sterile and the fertile tissue.

In *Riccia*, the sporophyte is simple without differentiation into foot, seta and capsule. In the sporophyte there is no sterile tissue excepting nurse cells or fore runners of elaters (in *R. crystallina* and *Oxymitra*).When the embryonic layers - endothecium and amphithecium are formed, the entire endothecium forms fertile tissue, the last generation of which is spores, while the amphithecium forms one layered capsule wall. In next ascending stage exemplified by *Corsinia*, there occurred sterilization of the basal part of the sporophyte to form a foot. This stage was succeeded by *Targionia* where sterile region consisted of foot and seta.The embryo divides in this genus and two fundamental tissues endothecium and amphithecium are formed. The amphithecium forms the capsule wall while the entire sporogenous tissue is derived from the endothecium. From this simple spherical sporophyte arises the sporophyte of *Marchantia*, in which sterile tissue is more pronounced. There is formation of seta and foot, both sterile tissues arising from the hypobasal cell. The epibasal cell gives rise to the capsule, the fertile region. Further sterility occurs when the sporogenous tissue derived from the endothecium forms two types of cells, the fertile - spore mother cells and the sterile - elater mother cells. In *Pellia*, the next member in liverworts, the hypobasal cell does not take part in formation of the sporophyte which in entirety is derived from the epibasal cell. Thus, the sterile regions – foot and seta arise from the same cell which is also involved in capsule formation. The sporogenous tissue derived from the endothecium is used up in formation of elaters, elaterophore and spores. This further adds to the sterility of the capsule region. Amphithecium however, is invoved in capsule wall formation.

Anthoceros, a hornwort with its unique sporophyte is example of further sterilization of the fertile tissue. The entire endothecium is devoted to a central sterile columella formation and the fertile tissue is shared with capsule wall formation from the amphithecium. The sporogenous tissue besides forming spores is consumed in formation of pseudoelaters, the sterile structures.

In mosses highest degree of sterilization is encountered. The sporophyte is differentiated into foot, seta, and capsule with opercular region. Not only seta and foot are the sterile tissues but also the opercular region with its peristome is sterile. The amphithecium goes into the formation of capsule wall which is elaborate. A thin sporogenous sac is formed from the outermost layer of endothecium (not so in *Sphagnum*) as rest of it forms columella. The archesporium which is the progenitor of the fertile tissue does not run in the opercular region and seta but is confined to theca.

The proponents of this theory (Bower, 1890) believe that the simple sporophyte gave rise to the complex one. It was further argued that this complexity was responsible for better dehiscence and dispersal mechanisms leading to greater survival value of the members.

**The other school of thought led by Kashyap, Church and Goebel believed that it was a retrogressive development of the sporophyte – from complex to simple, from moss-like to Riccia-like. Simplification occurred in dehiscence apparatus, in capsule wall and finally doing away with foot and seta.*

Liberation and Dispersal of Diaspores – Antherozoids, Gemmae and Spores (Plate VIII)

More than half the bryophytes are dioecious and it has helped in enhacing variability through outbreeding. This calls for an efficient dispersal of the gametes especially when the male and female plants are not close enough and when the antherozoids cannot travel a distance greater than a few centimetres. For bryophytes in which the antheridia are exposed in a 'splash cup' as in mosses, effective travelling distance is somewhat increased (Schofield, 1985). For that matter, in *Marchantia* also the receptacles help to raise antheridia above the thallus surface. The vegetative diaspore, gemmae in *Marchantia* are also dispersed to a distance by same mechanism. The dehiscence of the antheridium to release male gametes or antherozoids (*en masse*) occurs in presence of water. A highly coordinated mechanism occurs in which the antherozoid mass is dispersed farther so as to reach the archegonium.

In the simplest thalloid liverwort (such as *Riccia*), the mature antherozoids lie free, enclosed in a viscous fluid within the antheridium. The antheridium opens at the apex (in most cases this end has opercular cells), and the antherozoid-containing mucilage oozes out of the antheridium onto the thallus. At the air-water interface, the antherozoids spread apart rapidly forming a layer on the surface in which they are regularly spread apart. This spread is correlated with the presence of fat in the antherozoid mass. This fat lowers the surface tension and causes the spreading, in cases like *Sphagnum* and some liverworts. Absence of spreading could be correlated with absence of fats. Due to spreading, antherozoids are carried rapidly to the archegonium in the neighbourhood and they are then attracted chemotactically to the egg accomplishing fertilization.

In more complex situations, the dispersal may be explosive as in *Marchantia polymorpha*, *Preissia quadrata* and *Reboulia hemispherica*. In these species, the antherozoid mass is shot 2 to 5 cm and in *Monoclea forsteri*, it can be shot to even 8 cm while in *Conocephalum conicum*, antherozoids are dispersed to up to 12 cm into the air and in *Asterella californicum*, the explosive discharge is upto 20 cm. In all these species explosion is a result of swelling of mucilage cells on absorbing water which generates pressure. In the advanced bryophytes - the mosses, antheridia are produced in 'heads' which are excellent water traps, as in *Polytrichum juniperinum.* In such mosses splash cup mechanism is highly efficient and may carry the antherozoids up to 1.5 to 2 m away, though most splashes land much closer. Mosses which utilize the splash cup mechanism generally have longer fertilization ranges than mosses without splash cups.

In another group of bryophytes, the hornworts, antheridia are contained within embedded chambers in the thallus. Before maturity there's a narrow "roof" between the chamber and the outside world. When an antheridium matures the swelling slime content in the "roof" cells causes them to roll out, thereby creating a channel between the antheridial chamber and the outside. When the roof cells in a mature antheridium break down, the antherozoid mass oozes out onto the surrounding thallus.

Bryophyte antherozoids are typically coiled and have two whip-like flagella at one end. Experimental studies have shown that some bryophyte antherozoids can move at speeds of 1 to 2 tenths of a millimetre per second and maintain mobility for several hours.

Table 6.1 Bryophyte Review Sheet

Character	Phylum Marchantiophyta	Phylum Bryophyta "true mosses"	Phylum Anthocerotophyta , *Anthoceros*
Gametophyte characteristics	Thalloid Ventral surface with rhizoids & scales Air pores on dorsal surface Flat in cross section	Leafy gametophyte "stem" and "leaves" rhizoids anchor the plant 3-dimensional **Leafy gametophyte Lives in peat bog	Thalloid Ventral surface with slime pores *Nostoc* colonies in dorsal surface
Gametophyte special features	Gemma cups with gemmae for asexual reproduction Antheridiophores & archegoniophores *Dorsal furrow is where gametangia are located Unisexual or bisexual	Leafy gametophyte grows upright Male gametophyte with splash cup Protonema stage of gametophyte **Peat bog acidic Leaves with large, dead cells and small photosynthetic cells Can store 20X weight in liquid (water)	Dorsal surface appears dark green. Suborbicular mound of mucilage
Location of antheridia	On antheridiophore *In dorsal furrow	In splash cup of antheridial head of male gametophyte	On dorsal surface in chambers
Location of archegonia	On archegoniophore *In dorsal furrow	At top of female gametophyte (archegonial head)	On dorsal surface embedded, scattered
How fertilization occurs	Rain drop splashes antherzoids to top of archegoniophore *Lives in moist wet environment; antherozoids swim to archegonia	Rain drop splashes antherozoids from splash cup of male to top of female 'head'	Mucilage mound
Sporophyte characteristics	Consists of foot, seta, and capsule Attached to archegoniophore *Consists of simple sporophyte only Is inside calyptra	Consists of foot, seta, and capsule Attached to top of female gametophyte **Consists primarily of foot and capsule; seta represented by pseudopodium	Horn-like with involucre and basal meristematic region
How spore dispersal occurs	Seta elongates and pushes capsule out of calyptra; hygroscopic elaters cause spores to drop out *Sporogonium Thallus, calyptra and sporogonium must decay as there is no dehiscence mechanism	Calyptra drops off as capsule matures; operculum drops off when spores mature; hygroscopic peristome movement releases spores **Pseudopodium lifts capsule above gametophyte; capsule dries when spores mature; pressure builds within, operculum explodes off and spores released-shot gun mechanism	With hygroscopic twists of capsule wall and pseudo elaters.

Character	Phylum Marchantiophyta	Phylum Bryophyta "true mosses"	Phylum Anthocerotophyta , *Anthoceros*
Protonema	Mostly globose or thalloid, forming one bud, no gemmae	Filamentous, forming many buds, may produce gemmae	Globose
Special organelles	Complex oil bodies	Numerous choloroplasts, simple, small oil body	1-4 large chloroplasts, single plastid with pyrenoids
Gemmae	Common on leaves	Common on leaves, stems, rhizoids, protonema	Absent
Water conducting cells	Present only in a few simple thalloid forms	Present in both gametophyte and sporophyte	Primitive conduction
Stomata	Absent in both gametophyte and sporophyte, pores present	Present in sporophyte (Capsule)	Present in both sporophyte and gametophyte
Seta	Hyaline, elongating at maturity	Photosynthetic, rigid, persistent, elongating, pseudopodium	Absent
Calyptra	Ruptures and remains at base of seta, lacks influence on capsule shape	Ruptures and persists as cap influences capsule shape	Lacking sporophyte with involure
Capsule	Undifferentiated, jacket uni- or multistratose	Complex with operculum, theca and neck, wall multistratose	Undifferentiated, jacket multistratose
Sterile cells in capsule	Specially thickened elaters	Columella	Columella and pseudoelaters

* *Riccia*, ** *Sphagnum*

Evolution in Bryophytes: A New Perspective

The bryophytes, important components of virtually all terrestrial ecosystems, were among the earliest of land plants. How these plants diversified and evolved, has been reviewed in two recent books, an edited volume by Goffinet and Shaw (2009) and a text by Vanderpoorten and Goffinet (2009). Evidences from ultrastructural, biochemical, and molecular data support that the charophycean green algae (including the Charales, Coleochaetales, and other groups) are closely related to the embryophytes. Once the first land plants became successful in colonizing, the terrestrial ecosystems expanded and new groups of life appeared.

The first land plants (embryophytes) are believed to have appeared during the midPaleozoic era, between 480 and 450 million years ago. This was the result of emergence of aquatic algae from their habitats (Graham, 1993; Kenrick & Crane, 1997; Karol *et al.,* 2001; Waters, 2003; McCourt *et al.,* 2004). The transition was accompanied by various adaptations, including development of mechanical support for the plant body, formation of organs to exploit newly available resources, and the ability to resist new biotic and abiotic stresses. These adaptations involved changes in the composition of walls that surround the cells of all green plants (Niklas, 2004). Subsequently complex body plans and vascular tissues evolved that have allowed flowering plants to adapt to and survive in diverse terrestrial environments (Carpita & Gibeaut, 1993; Graham *et al.*, 2000; Cosgrove, 2005; Harris, 2005). The morphological, cytological, ultrastructural, biochemical evidences, molecular studies, and statistical sophisticated analyses, identify charophytes as the closest living relatives of the land plants.

However, a strong and abrupt evolutionary jump exists between these algae, where the diploid generation is represented only by the zygote, and in the land plants, the bryophytes, a pluricellular and organized structure represents sporophytic phase. In none of these living algae (or fossils, as far as it is known), an alternation of generations, *i.e.*, presence of a sporophyte, exists: all advanced charophyta have only the gametophyte. There is also a sudden appearance of multicellular sex organs with a sterile jacket of cells around the gametangium (antheridia and archegonia). The evolution of these gametangia and, in particular, of female sex organs or archegonia, is a fundamental event necessary for embryo development. It has been proposed that the last common ancestor of present-day land plants had a leafless, axial gametophyte bearing unicellular rhizoids and mucilage papillae. Vascular tissue, if present, was a central strand of perforate cells as in *Takakia* and *Haplomitrium*. Gametophyte axes bore terminal sporophytes consisting of a foot, a short seta and a capsule. As in modern liverworts, the seta probably elongated after spore maturation to enable the mature sporophyte to emerge from gametophytic involucres, and spore release was a one-time process. Meiosis was probably monoplastidic in ancestors, a condition considered plesiomorphic for extant land plants (Renzaglia *et al.*, 1993).

Chapter 7

Biological Interventions

Introduction

Bryophytes form a truly ancient lineage of plants. This has evidences from Charalean-like or Coleochaetalean-like algae which dominated lower Paleozoic era, and are believed to be the ancestors of land plants. Further, bryophytes are unique among land plants in having a dominant gametophyte phase, which makes them present opportunities for a broad array of research not readily undertaken in sporophyte-dominated organisms. The members can be manipulated experimentally and subjected to analytical testing in culture. This ease of culture has permitted numerous recent advances in areas of cell biology and physiology of this group of plants. Progress is also evident in deciphering the novel bioactive natural products of bryophytes. Understanding speciation and genetic diversity within species is another significant step in bryology. These plants have also enabled scientists to predict the consequences of environmental pressures created by the expanding human population and its agricultural and industrial base.

As Source of Bioactive Compounds

A great variety of lipophilic terpenoids, aromatic compounds, and acetogenins produced by liverworts have characteristic scents, pungency, and bitterness, and display a quite extraordinary array of bioactivities and medicinal properties (https:// www.researchgate.net/publication/228343350_Biologically_active_ compounds_ from_bryophytes). This has made bryophytes emerge as a potential biopharming tool for production of complex biopharmaceutiticals. As the field of medicines and drug discovery has expanded, the use of bryophytes for applied research with implications for human health has also broadened. Though bryophytes have been used as medicines in several cases, their utility as biopharmaceuticals is not investigated much. This is because bryophytes are mostly of minute size and it is difficult to identify diverse species of bryophytes. Bryophytes especially mosses and liverworts are the source of many biologically active novel compounds that could be used in pharmaceuticals. The occurrence of antibiotic substances in bryophytes has been well documented by botanists and microbiologists. Many species are known to possess compounds such as alkaloids (clavatoxine, clavatine, nicotine, lycopodine), polyphenolic acids (dihydrocaffeic) and flavonoids (apigenin, triterpenes *etc.*) but only a few are being investigated. With increasing demand for plant based medicines and rise in antibiotic resistant bacteria-search for new plant-based natural products is on. In a recent study, the biopharmaceutical company Greenovation Biotech Gmbh in Heilbronn, Germany, has made an effort to enhance the yield of recombinant proteins from moss. The moss, *Physcomitrella patens* has been successfully grown in a bioreactor with only water and minerals to nourish the moss. In presence of light and CO_2 (Greenovation) many complex proteins are produced in moss bioreactor. To tap the potential of bryophytes as source of novel drugs, bioprospecting of mosses and liverworts is suggested besides manipulating the plant system for the development of various novel therapeutic compounds.

As Research Tools

Bryophytes have been powerful experimental tools for the elucidation of complex biological processes. This is exemplified by investigations in *Sphaerocarpos*, a liverwort, the first plant in which sex chromosomes were discovered (Allen, 1917). With new techniques of molecular biology and bioinformatics coming up, analysis of mosses with modern molecular tools such as a knowledge of the genome via large-scale DNA sequencing, the ability to create transgenic individuals via transformation, and the capability to create gene knock-outs by homologous recombination research in bryophyte biology is being carried out. In mosses such as *Physcomitrella patens* (Cove *et al.*, 1997; Reski, 1999), *Funaria hygrometrica* (Schumaker & Dietrich, 1998), *Ceratodon purpureus* (Hofmann *et al.*, 1999), and *Tortula ruralis* (Wood *et al.*, 1999) focus is on molecular genetics. With its efficient gene targeting (Schaefer & Zryd, 1997; Strepp *et al.*, 1998), *P. patens* has been used to study the basic developmental mechanisms of land plants, particularly flowering plants (Prigge *et al.*, 2010). In recent years *P. patens* classified as dehydration tolerant (cannot withstand complete water loss) has emerged as model system to study mechanisms of abiotic stress adaptation, to identify stress-related genes and to characterize them functionally by reverse genetics approaches (Beike *et al.*, 2010).

Another upcoming field in bryophyte studies is the diversity of bryophytes which is challenged by profound global environmental changes. In Europe, for instance most of the endemic bryophytes are found in the Mediterranean region. This region however, is predicted to experience severe change of its climate in the next few decades (Gualdi *et al.*, 2013) and might witness a high risk of species extinctions (Thuiller *et al.,* 2005). As a consequence of global warming, therefore, significant losses in bryophyte diversity are expected. The areas harbouring large number of species such as boreal forests of higher latitudes, alpine biomes and at higher altitudes of tropical mountains will be most affected. There may be other regions around the globe with similar alarm. Therefore, improving understanding of the effect of rising temperatures on bryophytes is particularly important. It is equally important to explore ecophysiology of bryophytes in the changing environment.

Reasons for bryophytes as popular research tools are many. Compared to tracheophytes, bryophytes have no mechanical protection like bark or cuticle. Most bryophytes grow on forest ground in a close connection to several biodegrading destruents and protection against pathogens like fungi or bacteria is essential for surviving in such habitats. The plants therefore, exhibit antibacterial and antifungal properties (Merkuria *et al.,* 2005; Zhu *et al.,* 2006).

In contrast to most seed plants, bryophytes can be preserved deep-frozen in or over liquid nitrogen, a method optimized by Schulte and Reski (2004) for *Physcomitrella* mutants. Even after ten years deep frozen, moss mutants can be thawed and re-grown unchanged. **Recently, this technique has been used to establish the International Moss Stock Center (IMSC), based at the University of Freiburg, Germany and can be accessed via www.moss-stockcenter.org.**

Mosses are preferred model systems in many research areas for simple reasons. At least for 330 million years, the simple morphological structure of mosses is largely conserved (Hubers & Kerp, 2012). The mosses colonize an extensive spectrum of diverse habitats (Turetsky, 2003; Turetsky *et al.,* 2012) and occur in many extreme habitats such as the Antarctic tundra and deserts. It is an indication that many moss-specific survival mechanisms exist though the pathways remain unclear (Oliver *et al.,* 2005; Roads *et al.,* 2014). The resilience of mosses has been recently demonstrated with the successful regeneration of subglacial bryophytes following 400 years of ice entombment (Farge *et al.,* 2013) and by the regrowth of over 1500-year-old moss from the Antarctic permafrost (Roads *et al.,* 2014).

As Ecological Tools

Mosses are also important components of tropical systems (Gradstein *et al.,* 2001), boreal forests, and woodlands or unshaded habitats in temperate zones (Turetsky, 2003). Despite their small size, they are impactful on various ecosystems and are essential contributors to complex biological cycles. They contribute to the prevention of soil erosion, the development of microtopography, and the regulation of soil climate, water availability, and nitrogen fixation and carbon cycling. In forest ecosystems such as rain forests, mosses live in immediate vicinity to vascular plants, where they serve as water reservoir (Gradstein *et al.,* 2001) as well as bio-indicators for environmental changes and air pollution (Bates, 2009).

Hence, mosses represent an omnipresent organism group which deserve attention as attractive model systems regarding ecology as well as study of fundamental processes and evolution of land plants. Moss model species include, *e.g., Funaria hygrometrica, Physcomitrella patens, Ceratodon purpureus,* and peat moss (*Sphagnum*) species. The most developed moss model is of *P. patens,* from the temperate zones which can be found on soil exposed by falling water levels or on fields (Cove, 2005). It provides a fully sequenced genome (Rensing *et al.,* 2008), many resources for -omics techniques, a high amenability to microscopy, and a valuable experimental platform for comparative studies with other model organisms, as, *e.g.,* trees. Therefore, the *Physcomitrella* genome was designated, like poplar, as *"JGI plant flagship genome".

As Evolutionary Group

Since liverworts represent earliest diverging group of land plants, many fundamental features of land plants first appeared in the gametophytes of bryophytes. These were believed then co-opted to the sporophytes in vascular plants (Ligrone *et al.,* 2012, Pires & Dolan, 2012). Therefore, to determine the fundamental mechanisms and principles common to land plants, the molecular biology of basal plant lineages has been investigated using modern tools. In view of its critical evolutionary position, *Marchantia polymorpha* became part of community Sequencing Program at the Joint Genome Institute (DOE-JGI: http:// jgi.doe.gov/ why- sequence-a-liverwort/). The whole-genome analysis of the plant was initiated under which it was revealed that most of the genes that regulate growth and development in other land plants were conserved, but they showed less redundancy, in *M. polymorpha* genome. It is to be understood that *Marchantia* cultured cells were the first to be fully sequenced in plants. The Y chromosome, as well as a part of the X chromosome, of *M. polymorpha* was also sequenced which has provided insights into the evolution of sex chromosomes in organisms with haploid genomes (Yamato *et al.,* 2007).

Coming to hornworts, they have been at the key position in evolution of land plants. Ferns, a group of early land dwellers dominated the botanical world for hundreds of millions of years, between the Devonian, about 360m years ago, and the rise, about 120 m years ago in the Cretaceous. Most ferns could not stand the competition from other plants and were driven to extinction. However, one group of these ferns survived competition and diversified. Almost all ferns (extant sp), now alive are its descendants. These ferns have an ability to live in the shadows of their competitors. Intriguingly, these successful ferns also evolved a new kind of light-sensing protein known as a neochrome. Due to this protein, ferns became sensitive to dim levels of light. According to researchers, these neochromes may have enabled ferns to thrive on shady forest floors. A theory has been proposed

---

**Discussion with the JGI (Joint Genome Institute) Plant Genome Advisory Committee and DOE (Department of Energy), led to an understanding that a set of JGI plant genomes that are the most important to DOE mission and plant science have been designated as JGI Plant Flagship Genomes(http:// jgi.doe.gov/our-science/science-programs/plant-genomics/plant-flagship-genomes/).*

that during evolution ferns got a new gene for sensing light from a moss-like plant called hornwort. The investigations revealed that this gene did not gradually evolve in ancient ferns but instead, a single lineage of ferns picked up the neochrome gene from hornworts about 180 million years ago when a hornwort and a fern grew in close proximity. (Plants That Practice Genetic Engineering - The New York Times; https://www.nytimes.com/2014/04/17/./plants-that-practice-geneticengineering. html).

Conclusions

Today bryophytes have increasing importance as sources of valuable substances for biomedical applications. Though the life cycles of vascular plants and bryophytes differ to a large extent, there is also a great evolutionary distance between the two. Researches on bryophytes have indicated that some of the regulatory pathways found in angiosperms also exist in these archegoniates. The evolutionary studies support a model of plant evolution whereby several features found in vascular plants were developed by harnessing gene regulatory networks that were present in the common ancestor of bryophytes and vascular plants (Pires & Dolan, 2012, Kubota *et al.,* 2014, Flores-Sandoval *et al.,* 2015, Kato *et al.,* 2015). To elucidate these pathways studies have largely concentrated on model system *Physcomitrella patens*. In recent years liverwort *M. polymorpha* because of its critical phylogenetic position is accepted to have substantial potential as a model system for plant biology in the evolution of land plants. With its conserved, less complex, developmental pattern and responses to plant growth factors and environmental stimuli, it is part of several projects working to reveal the evolution and diversity of regulatory systems in land plants (Ishizaki *et al.,* 2016).

Chapter 8

Collection and Processing of Bryophytes

On-field collections make an important part of studies on bryophytes. Some common liverworts and mosses can be collected and identified early based on their reproductive structures **(Plates IX and X).**

To begin with the field collections, it must be remembered that bryophytes are small plants and have thalli which are fragile that can get damaged easily if, not sampled carefully. Also while searching for liverworts and mosses, the best places would be near some damp humid ground or water body. However, some genera grow on dry rocks and wood also. The good quality specimen should ideally consist of both the gametophyte and the sporophyte generations. Most often, the gametophyte generation is present and easily identifiable in the field, but it may not always bear a sporophyte, an important character in moss identification. Therefore, trips to the field should be planned when the reproductive stages are also expected. The months of October to December are good periods and collection of viable, actively growing, material will facilitate a more reliable identification. When collecting a sample, it is also important to collect a "representative" sample, which is valuable for making observations of the general habit of the plant and to adequately sample any variation present within a population.

Bryophytes are among the easiest plants to collect (Buck & Thiers, 1996), since they lack roots, they can often be readily collected by hand. Some species which grow closely attached to their substrate should be scratched using a knife. There is no pressing or mounting involved as in ferns and gymnosperms, all one has to do is to dry them, and put them in a permanent packet with good collection data or a wet specimen can be prepared. Though these tips might appear simple but one tends to ignore these and collection may not be worth the effort.

A few on-field tips to collect bryophyte samples for herbarium, are provided:

Collection

The tips may appear simple but if ignored, collection may not be worth preserving.

1. Collect groups of entire plants

- All parts of a plant help in identification and minimum injury should be caused while collecting the specimens.
- A knife or a small chisel is used to lift plants from rocky or woody substrates without leaving their bases behind.
- A small sample of substrate is collected with the plants if they are too firmly attached to their substrates and is difficult to lift them in one piece (generally collecting large quantities of soil or litter with the plants should be avoided).
- Plants with sporophytes (capsules) - even old ones – are more useful for identification than those without.
- Each sample should ideally be about the size of the palm.

** If necessary, a smaller specimen can be used for identification. It is important to remember that for species to remain at the site for future, never collect a sample if it will remove most or all of the species from the site. It is extremely important to remember that species extinction has become fast with ecosystem degradation.*

2. Each sample should have a single dominant species

** Bryophytes often grow intermixed, therefore separate collections for each species should be made. For example, if species A is mixed with B, attempt should be made to collect a sample of each species A and species B in separate bags.*

Paper is used instead of plastic because bryophytes are preserved best in a dry state. Number 2 paper bags with 210 x 200 mm (outer) 195 x 200 mm (inner) dimensions are recommended for their convenient size. However, any paper container can be used as long as the collection is secure. Such bags can even be made from newspaper. The top of each bag should be folded over so that the sample doesn't fall out. Students can prepare bags from newspaper before they go to field.

- Ideally, each sample should be packaged separately rather than putting several collections together in one bag. This is preferred because on drying, bryophyte samples become brittle and difficult to separate from one another without inflicting damage to the specimen.
- Some samples that form solid cushions when moist can fall apart when dry and mix together with other species in the same bag. Further more,

mosses in over-stuffed bags may dry too slowly and the collection gets prone to microbial attack.

3. Each bag should be legibly labeled

- Microhabitat information such as substrate (aspen base, calcareous boulder), the level of light (sunny, shaded, partly-shaded), the level of moisture (dry, mesic, wet) should be marked on each paper bag.
- Codes for wet, for dry, shaded sand and other habitat specification could be used for these microhabitat characters.
- Each bag should also be labeled with information that links it to the site of collection and to the person who made the collection. It might also include latitude.
- To save field time, some collectors use pre-label bags with consecutive numbers, and then associate the collection numbers with site information written down in a separate field book. This requires a little extra attention since the pre-labeled bags should ideally be used in numerical order otherwise error chances to mix up the information increase.
- Carry a separate bag for collections. A large cloth sack is ideal as it allows damp samples to breathe instead of soaking and disintegrating the paper collection bags, and will also hold together better than plastic when dragged through the sampling sites.
- However, in pouring rain which is unpredictable on hills, plastic bags may be preferred. The whole sack can be placed in a backpack or strapped to the outside. This arrangement secures the samples which otherwise might get lost when other equipment is pulled from the pack.

4. Dry samples as soon as possible

- To prevent mildew or bacterial damage, paper bags containing samples should be spread out in a single layer in a dry place right after the collecting trip.
- The samples can be spread out on newspaper which can be put in press or under some weight. These newspapers should be changed initially daily, then on alternate days and finally after a week or so depending upon the wetness of the specimens **(Fig. 8).**
- If the field trip lasts for several days, samples should be spread on dry ground in the sun. They should be shifted periodically until their contents are completely dry.
- Dry samples can be stored almost indefinitely in their bags, in a dry place.
- For wet specimens, collection can be put in vials containing 4% formalin.

Remember – A specimen without field data (at least locality and date) is of very little scientific value.

Once the specimens have dried, they can either be put as herbarium **(Fig. 8)** or processed for section and staining. This helps us study the anatomical details of the part preserved.

The material can be simply mounted on the slide, stained and observed as whole mount or can be dissected as preferred for genera like *Anthoceros*. However, in most cases it is sectioned, stained and a temporary mount made.

A

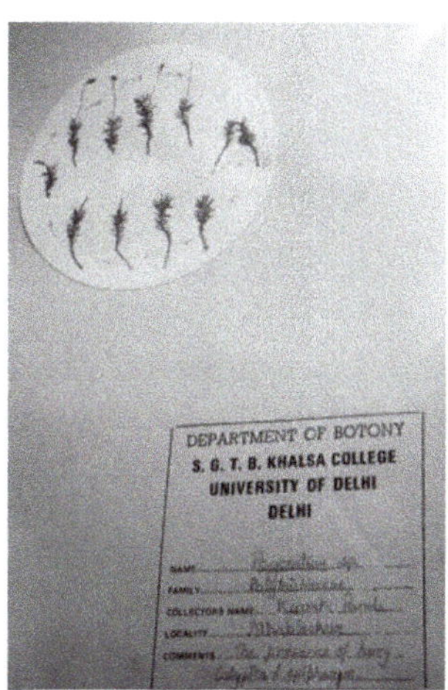

B

Figure 8*A&B* On-field collection requires a press *(A)*. Herbarium sheet to show the label with field information *(B)*.

Sectioning and Staining

- A pith piece is taken and split longitudinally, or held between thumb and fingers and cut in halves with a sharp knife/blade. In certain cases, groove may be made instead so that the delicate material could be placed securely in position and held firmly.
- Having placed the specimen in position between the two pieces of pith, trim off the end and cut thin sections from it, judging the thickness of each section of the specimen by the opacity or transparency of the pith sections cut along with it.
- A good number of sections are cut, transferred to weak alcohol in a watch glass or petridish and picked out.
- The sections are stained in safranin and mounted in glycerine.

HOYER'S Solution as a Rapid permanent mounting medium for bryophytes by Lewis Anderson (1954).

Hoyer's solution is a rapid permanent mounting medium which has been in general use for many years. Solution has been tested on considerable variety of mosses and liverworts. Formula to prepare :

- Take distilled water 50cc, gum Arabic 30gms, chloral hydrate 200gms, glycerine 20cc. The ingredients are mixed in above order at room temperature.
- As gum Arabic goes slowly into the solution therefore, one can use USP Flakes (United States Pharmacopeia, if available).
- Solution is allowed to stand for several hours and stored thereafter, in air tight bottles. The final mix should appear clear with only a faint tinge of yellow.
- Plants can be transferred to Hoyer's solution directly from water.
- Cover glass is lowered and slide sealed. Hoyer's solution is an extremely effective clearing agent.

** A lactophenol gel made with methyl cellulose is described as a combination clearing and mounting medium for bryophytes with delicate cell wall structure collapsing in other media such as Hoyer's medium.*

Colour Plates

PLATE I: *Marchantia*

1. V.S. Thallus to show organization of tissues into two distinct zones – photosynthetic (ph) and storage (st) zone. Also seen are the air pores (arrow).

2. Specimen - Thallus with gemma cups (arrows). The cups are raised structures with frilly margin and a base with gemmae.

3. V.S. Thallus through gemma cup. Note the tissue forming the cup is built on same plan as the vegetative thallus, with a photosynthetic and a storage zone. Several gemmae (ge) arise from the floor of the gemma cup.

4. W.M. Single gemma with two distinct points of apical growth. The central tissue disintegrates and apical notch (arrows) cuts off cells that give rise to a new plant.

5. Specimen - Male plant with antheridiophores.

6. L.S. Antheridiophore shows chambered structures with a single antheridium (an) in each. The air pores (arrows), rhizoids (rz) and scales present on the stalk indicate that the construction plan is same as that of vegetative thallus.

7. Specimen - Female thallus with archegoniophores.

8. L.S. Archegoniophore to show peltate disc with youngest archegonium towards the stalk. The archegonia are inverted and possess various protective layers (inset).

9. L.S. Archegoniophore post-fertilization, to show developing sporophytes (sp) which remain hanging on the peltate disc.

10. *A&B*L.S. Archegoniophore with young sporophyte magnified to show short seta (se) and the spherical capsule (ca). The capsule has a wall and is covered with perigynium (arrow) and calyptra. In (B), the developing sporophyte shows an elongated seta (se).

11. L.S. Archegoniophore bearing fully developed sporophyte with distinct foot embedded in placental tissue, elongated seta and oblong capsule with elaters and spores. The protective coverings have disintegrated with the elongation of the seta.

12. L.S. Archegoniophore. Diagrammatic sketch to show protective layers – calyptra (1), perigynium (2) and perichaetium (3).

Plate II: *Riccia*

1. **Specimen – Habit. Plants grow in clusters forming rosettes.**

2. **W.M. Thallus. Ventral view of a distinctly dichotomously branched thallus with apical notch (arrows) and profuse growth of rhizoids along the midrib.**

3. **V.S. Thallus shows photosynthetic (ph) and storage zones (st). Also seen are rhizoids (rz) and scales on the lower surface.**

4. **V.S. Thallus showing archegonium (arrow) deeply placed in the dorsal surface.**

5. **W.M. Thallus to show sporophytes along the midrib, where youngest (arrow) is present towards the apical notch while the mature ones (double arrows) are towards the base.**

6. **V.S. Thallus to show sporophytes at different stages of development.**

7. ***A&B* V.S. Thallus. (A) - with sporophyte at spore tetrad stage. (B) - at spore stage. The remnants of calyptra are seen as the mature sporophyte is covered with only single layered wall. Note the sporophyte is simple without distinction into foot, seta and capsule.**

Plate III: *Pellia* and *Porella*

1. ***Pellia* – Specimen. Thallus bearing sporophytes with involucre (arrow) at the base.**

2. **V.S. Thallus to show simple organization without any distinction into photosynthetic and storage zones as seen in Marchantiales. Rhizoids (rz) are seen to arise from lower surface.**

3. ***A&B* V.S. Thallus. Diagrammatic sketch showing a monoecious condition with archegonia (arrow), a little behind growing tip and antheridia further away from the archegonia. In (B), fertilized archegonium has formed well a differentiated sporophyte (sp).**

4. ***A&B* L.S. Sporophyte with a distinct foot, seta (se) and capsule (ca). Also seen are elaterophore (arrow), spores and elaters. Developing sporophyte is still enclosed in the involucre (double arrow). Also seen is the young archegonium with a zygote (arrow). In (B) the ruptured involucre (arrow) stays behind as the sporophyte elongates.**

5-7 *Porella* – *5*- Leafy axis with dorsal leaves closely overlapping in an *incubous arrangement* i.e. the lower edge of each leaf is covered by the upper edge of the next leaf below. (6) - The monoecious axis shows a single globose antheridium (an) in the axil of the leaves and terminal archegonia (ar) in groups. *7*- Antheridium is magnified to show a globose structure on a long stalk (arrow). The antheridial wall is two celled.

8. **W.M. Leafy axis showing a young sporophyte.**

9. **L.S. Sporophyte with protective perianth (pe) and involucre (in). The foot is indistinct, seta elongate and capsule is spherical. Also seen is the underdeveloped archegonium (arrow).**

Plate IV: *Sphagnum*

1. **Plants with comal tufts bearing sporophytes.**

2. **T.S. Stem to show cortex followed by a prosenchymatous region and the central cylinder (cc).**

3. **W.M. Leafy axis to show retort cells (arrows) which help plant store water.**

4. **W.M. Leaves with hyaline cells showing horizontal bars which prevent collapse of these cells. Narrow photosynthetic cells are also seen.**

5. **L.S. Sporophyte with pseudopodium (pd), a constriction in neck (arrow) and globose capsule enclosing spore sac and columella (co). Also seen is vaginula (v), the region of stalk with foot of sporophyte embedded in it.**

6. **Stages in capsule dehiscence. Diagrammatic representation of capsule dehiscence marked by change in shape, from spherical to cylindrical.**

7. ***A&B* Products formed from *Sphagnum* based on the antiseptic properties of the moss.**

Plate V: *Polytrichum*

1. **Specimen – Leafy gametophore bearing a sporophyte. Note the capsule is straight, held high up in the air by a slender seta (arrow).**

2. **Specimen - Hairy cap moss gets its name from the hairy calyptra.**

3. **Specimen - The leaves towards the apex of the stem are green and closely placed.**

4. **W.M. Leaf shows narrow wing and expanded sheath. Note the 'hinge' (arrow) where the lower portion of leaf expands to form a sheath.**

5. **W.M. Rhizoids which are branched and have oblique septa.**

6. ***A&B* V.S. Leaf with narrow wings and a well-defined midrib. In (*B*) portion magnified to show photosynthetic lamellae whose terminal cell bifurcates (arrow) and therefore, is characteristic of species.**

7. **T.S. Stem with spirally arranged leaves also cut in the same plane.**

8. **T.S. Rhizome to show water conducting hydrome and starch storing leptome.**

9. **W.M. Antheridial head. Note, the red colour of perigonial leaves impart colour to male branch.**

10. **L.S. Sporophyte with distinct zones-seta (1), apophysis (2), theca (3) and operculum (4).**

11. **W.M. Nematodontous peristome with a fixed ring of teeth. Also seen are spores.**

Plate VI: *Funaria*

1. Specimen – Habit. The moss grows in tufts and is characterized by obliquely placed capsule.
2. W.M. Plant. The gametophore remains attached to the substratum by rhizoids.
3. W.M. Gametophore with leaves closely placed towards the apical portion of the stem.
4. W.M. Leaf shows midrib (costa) and the single celled wings whose cells are rich in plastids.
5. V.S. Leaf to show multicelled midrib and a single celled wing.
6. W.M. Antheridial head.
7. W.M. Archegonial head.
8. L.S. Female head to show archegonia with long necks. Also seen are paraphyses (arrows).
9. L.S. Sporophyte - Diagrammatic sketch with distinct opercular region, theca, apophysis and seta.
10. W.M. Rim (r) and annulus (a).
11. W.M. Outer peristome is ornamented with horizontal bars.
12. Photograph - The outer peristome bends outwards when the air is dry and facilitates spore dispersal while in moist conditions the peristome incurves closing the mouth.
13. W.M. Protonema and leafy gametophore.

Plate VII: *Anthoceros*

1. Specimen - Plant in its natural habitat. The upright sporophyte gives it the name – 'Hornwort'.
2. W.M. Thallus ventral view to show rhizoids.
3. *A&B* V.S. Thallus to show simple construction. In (*B*), portion is enlarged to show distinct plastid (arrow) in each cell.
4. V.S. Thallus to show slime cavity.
5. W.M. Thallus to show tubers. Note thick wall of the tuber.
6. W.M. Thallus to show antheridia (arrows). Also seen are *Nostoc* colonies (double arrows).
7. V.S. Thallus through antheridial chamber shows a primary antheridium with two secondary antheridia.

8. V.S. Thallus with archegonium, endogenous in origin.

9. L.S. Thallus through the foot region (f) of the sporophyte. Also seen is the slime cavity (arrow).

10. L.S. Young sporophyte, covered by the involucre (in).

11. T.S. Sporophyte to show columella and spore tetrads along with pseudoelaters. The capsule wall is multilayered.

12. L.S. Sporophyte with columella (1), pseudoelaters (2) and spores (3). Note multistratose wall.

13. W.M. Pseudoelaters and spores magnified.

Plate VIII. *Dispersal mechanisms in bryophytes*

1. *A&B Marchantia* with gemma cups. (*B*) - Line diagram to explain 'splash cup' phenomenon.

2. Moss - Release of antherozoid mass from antheridial head; the antherozoids land on the archegonial neck by 'splash cup' phenomenon. The female branches are always placed lower than the male branches in such mosses, facilitating fertilization.

3. *Sphagnum* – Change in shape of capsule from spherical to cylindrical facilitates spore release.

4. *A&B Polytrichum* – First visual change in capsule dehiscence and spore release is throwing away of the operculum and exposing the peristome. (*B*) - Peristome and epiphragm. The ridge and spur structures on inner side of peristome teeth alternate with sacculion edge and lower surface of epiphragm. For details visit: http://www.helsinki.fi/~jhyvonen/bell&hyvonen2010AJB.pdf.

5. *Funaria* – shows a dome, formed by the peristomial rings and no spores are released (upper half of diagram). Later, the outer peristome outcurves making way for spores that are sifted through the inner peristome(lowerhalf of diagram).

6. *A-C Marchantia.* The sporophytes hang on the lower side of the archegoniophore as shown in (*A*). The elaters which are hygroscopic twist and help in shedding of the spores (*B*). Emergence of the sporophyte from the protective coverings also plays an important role in dehiscence of capsule and dispersal of spores (*C*).

7. *A-C. Pellia.* The capsule bears thin lines of dehiscence (*A*) along which the capsule dehisces at maturity. The valves roll back as in (*B*) . The spores with elaters are exposed and spores are dispersed by combined effort of elaters and wind (*C*). Also seen are the fully rolled back valves.

8. *A&B Anthoceros.* The sporophyte is unique with continuous spore production and release over a long period of growth.

Plate IX: *Some common liverworts for field study*

1. ***Conocephalum conicum*** **– great scented liverwort. The thalli are very strong-smelling, with purplish margins; a dark green, leathery surface; flat and smooth. There is a set of lines running along the thalli's surface. The air pores, which are found between the lines, are more conspicuous.**

2. ***Lunularia cruciata*** **- The crescent-cup liverwort. The name is from Latin '*luna*' meaning moon which refers to the moon-shaped gemma cups.**

3. ***Mannia triandra*** **- The species is sensitive to successional changes of the vegetation related to the overgrowing of open habitats with scrubs, afforestation and overgrazing and is included in the Red Data Book of European Bryophytes as "Rare".**

4. ***Reboulia hemisphaerica*** **– Archegonia are borne on the receptacles.**

5. ***Peliia epiphylla*** **– Reproductive plants with sporophytes.**

6. ***Frullania tamarisci*** **– The millipede liverwort. Two lobes of the leaves are folded against each other while under lobes are small and helmet-shaped.**

7. ***Scapania Undulata*** **- Has rounded leaf lobes appressed to each other, a straight keel, a front lobe that does not run down onto the stem and a back lobe that does.**

8. ***Plagiochasma rupestre*** **– Sex organs are borne one after other in succession.**

9. ***Asterella californica*** **-The receptacles are rounded, with four lobes each bearing a single sporophyte sheathed by a white tattered skirt.**

10. ***Blasia pusilla*** **- The plants produce two types of gemmae: loose clusters of star-shaped gemmae on the thallus surface, and spherical gemmae borne in beaked, flask-like receptacles.**

Plate X: *Some common mosses for field study*

1. ***Barbula unguiculata*** **– The common beard moss. The slender setae are erect or ascending, long, terete, and hairless, are usually orange to red at maturity. The capsules with membranous hoods or calyptra, are nearly cylindrical in shape, but slightly broader at the base than the apex and orange, red, or brown at maturity. The operculum has long narrow beaks, often tilted to one side.**

2. ***Pogonatum aloides*** **– Spike moss. One of the most conspicuous features of the plant is the delicate whitish epiphragm that covers the mouth of the ripe capsule after the lid has fallen. These discs of opaque white catch the eye from a long distance.**

3. ***Bryum argenteum*** **- Mature capsules are cylindrical-obovoid in shape, hairless, and pink, red, or reddish brown. They droop down from the apices of their stalks. The small operculum of the capsule has a short-conical dome-like shape. The calyptra covers only the operculum and uppermost body of their capsules; the hoods are membranous, strongly beaked, and early-deciduous.**

4. ***Tortula muralis*** **– Sporophytes are slender with a conical and pointed tip. Calyptra symmetrical; splitting down one side.**

5. ***Encalypta*** **- Capsules are erect, with cylindrical theca, occasionally furrowed longitudinally or spirally; neck is usually indistinct and capsules appear massive, glossy, crimson-red, with conical operculum that is short- to long-rostrate.**

6. ***Buxbaumia*** **(bug moss) - The capsules orient themselves so that the mouth is pointed towards the highest light intensity. Puffing the spores in that direction would increase their chances of clearing surrounding obstacles and dispersing further away.**

7. ***Schistidium rivularis*** **– Sporophyte showing operculum with separated columella still attached. This continued attachment of columella is unusual.**

8. ***Hennediella heimii*** **- Dehisced capsule with operculum attached via the columella.**

PLATE I: *Marchantia*

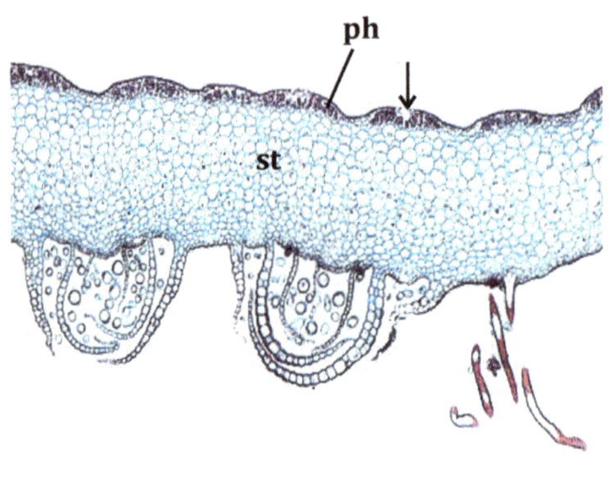

1

2

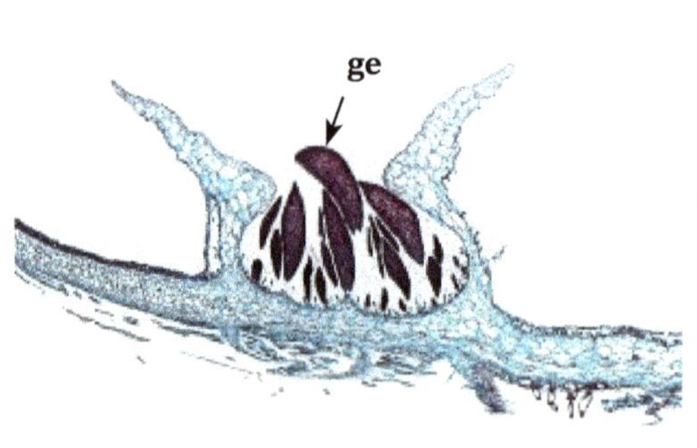

3

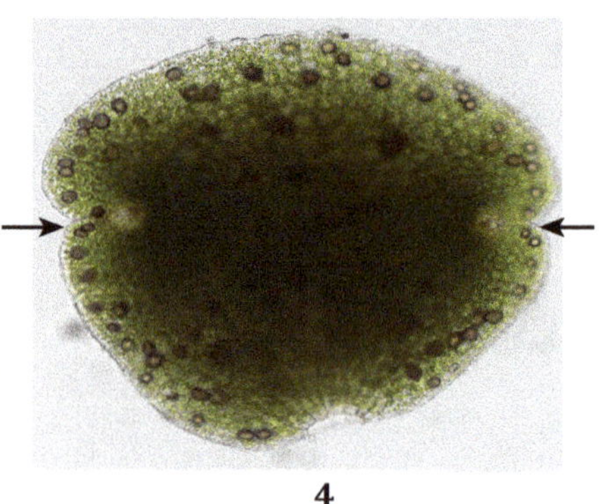

4

5

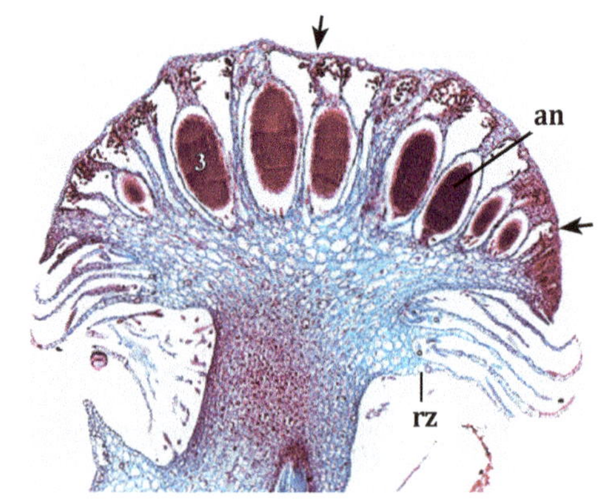

6

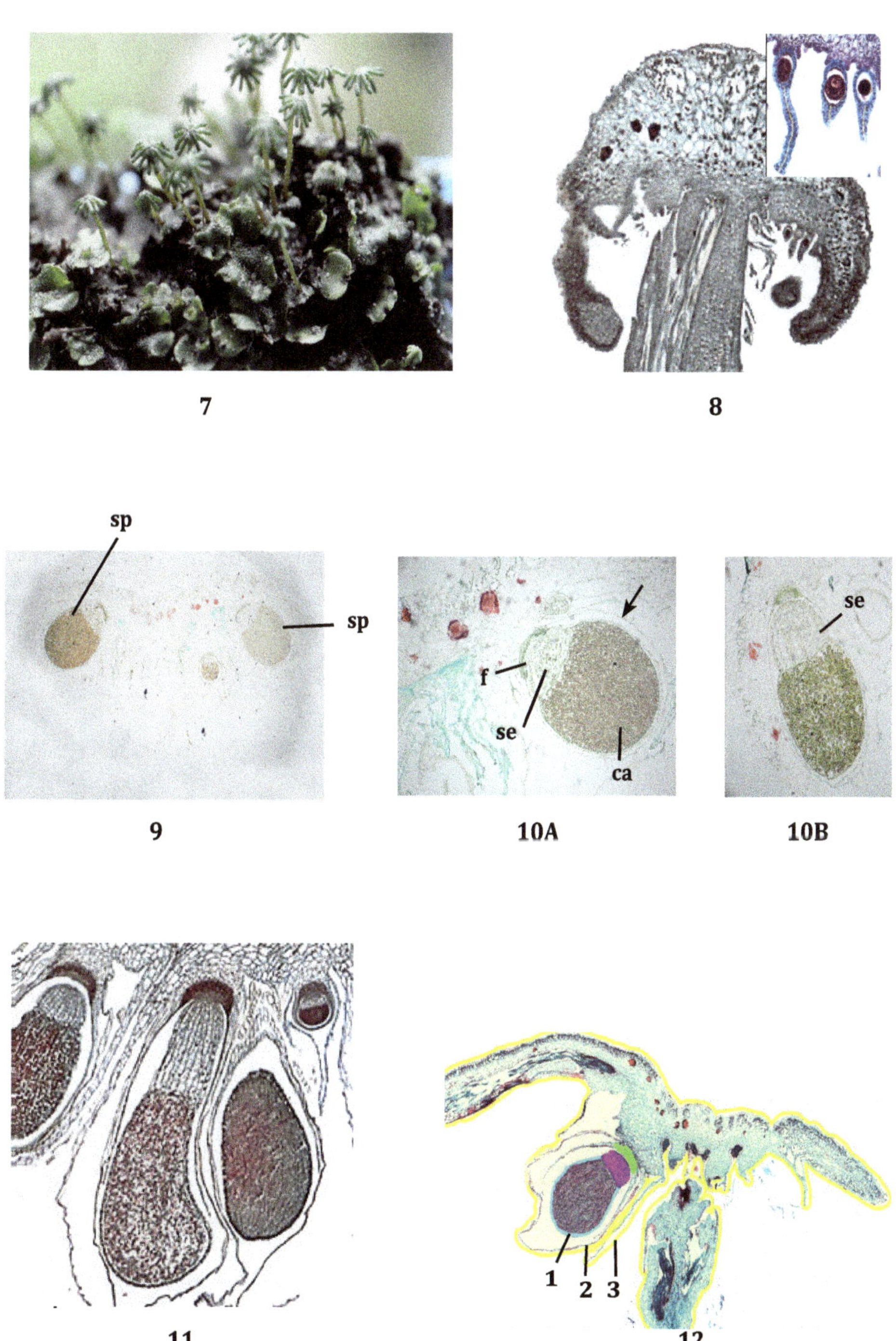

7

8

9

10A

10B

11

12

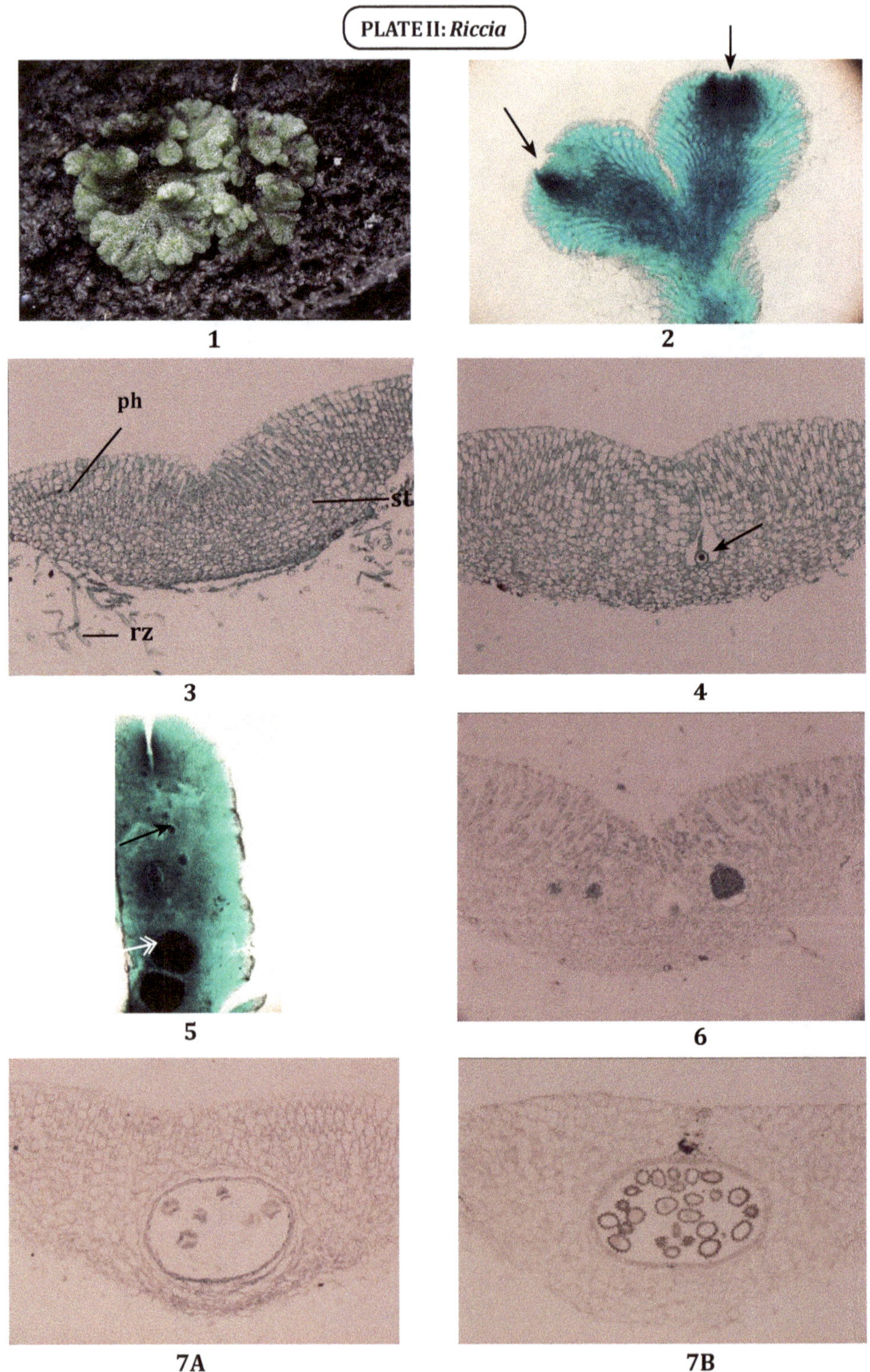
PLATE II: *Riccia*
1
2
ph
st
rz
3
4
5
6
7A
7B

PLATE III:Pellia and Porella

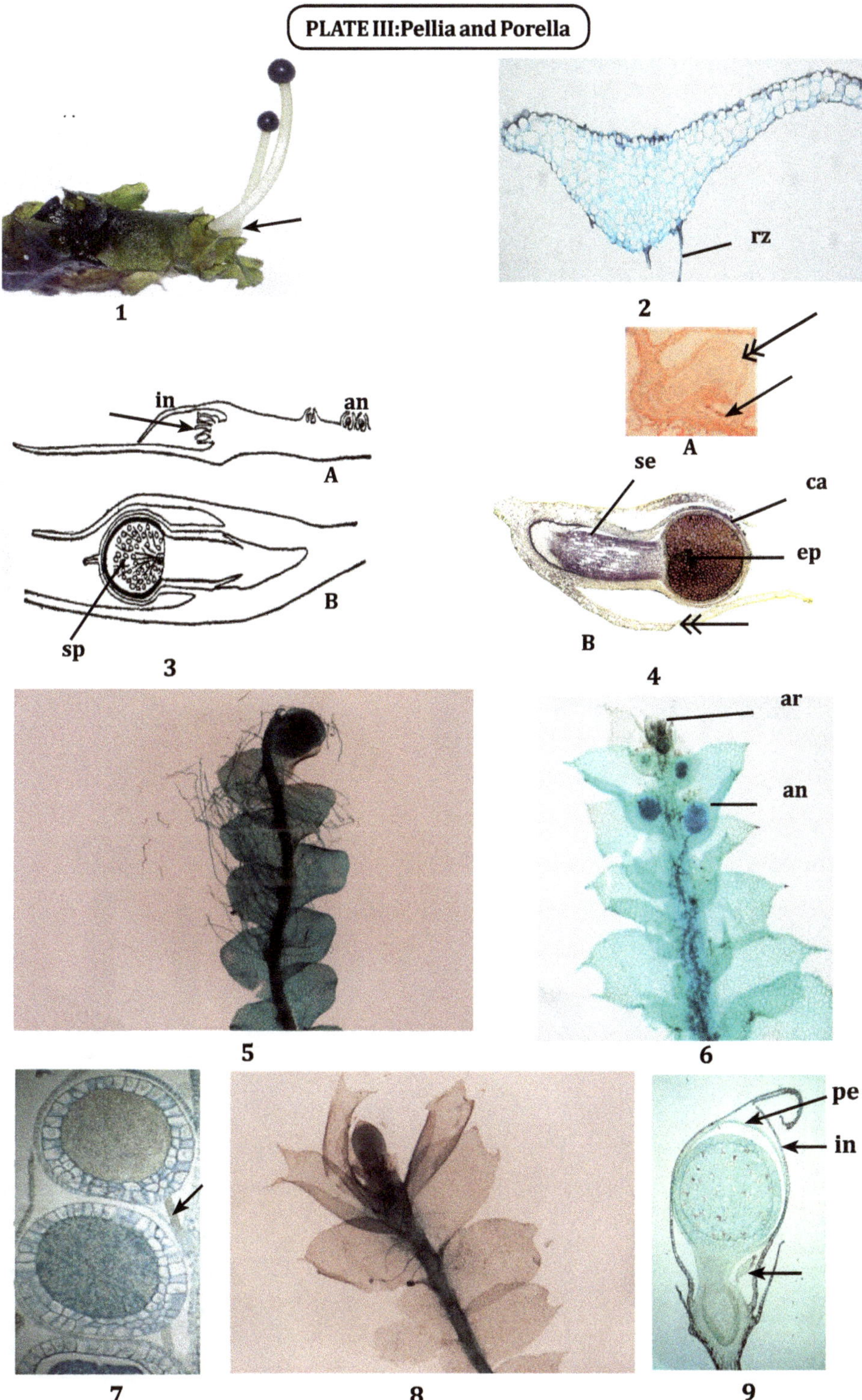

1 2 3 4 5 6 7 8 9

PLATE IV: *Sphagnum*

1

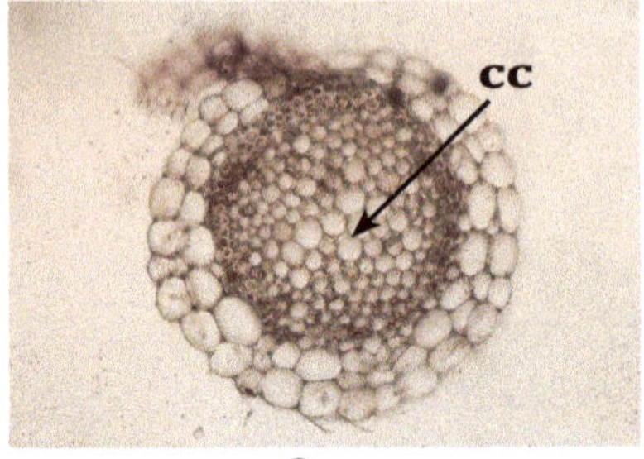

2

3

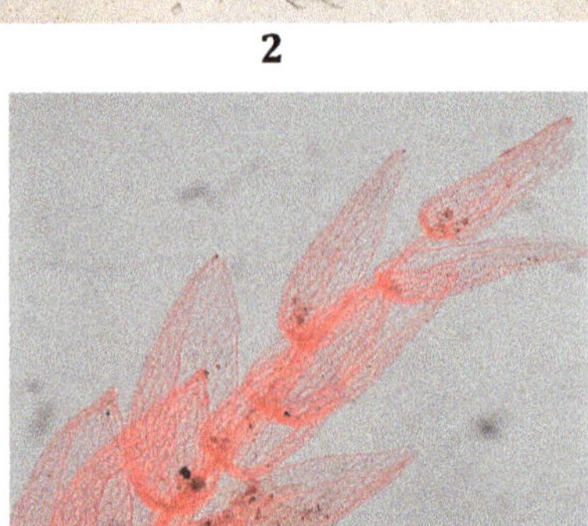

4

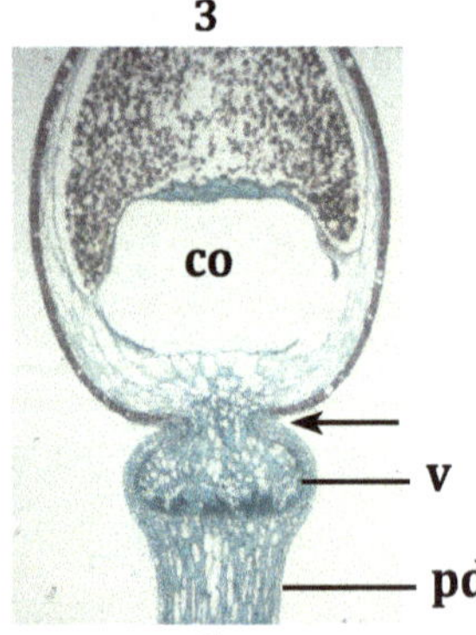

5

6

7A

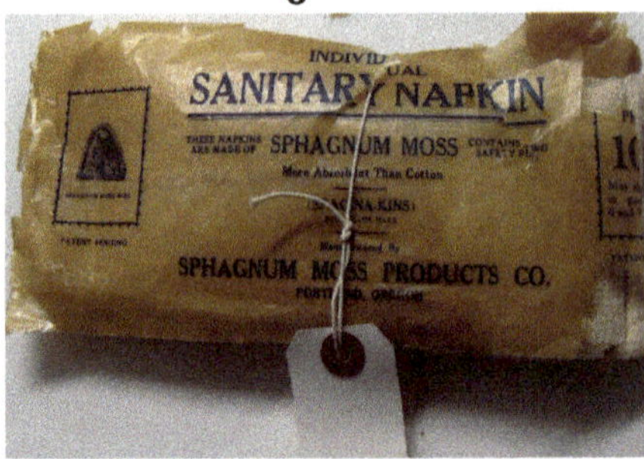

7B

PLATE V: *Polytrichum*

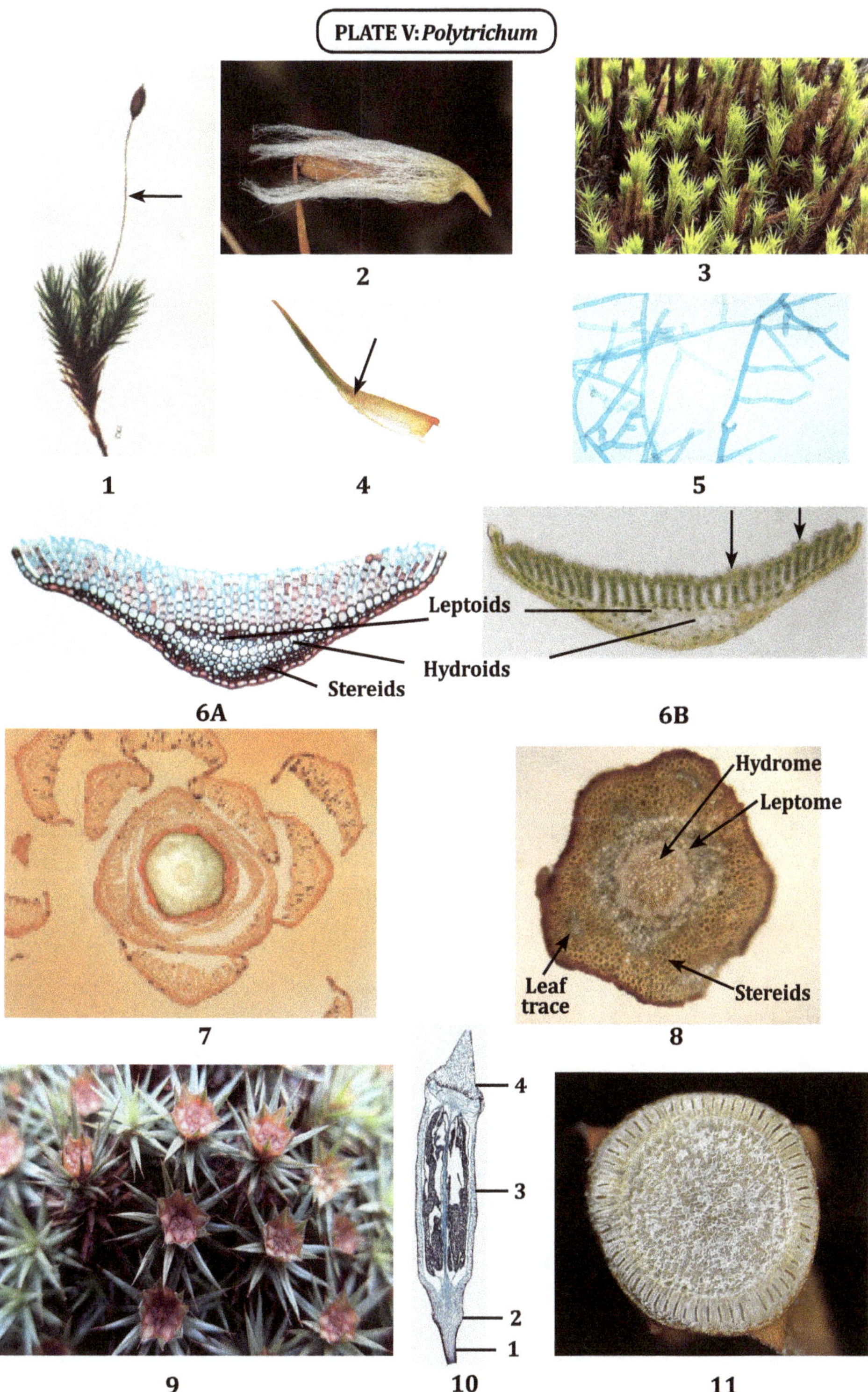

PLATE VI: *Funaria*

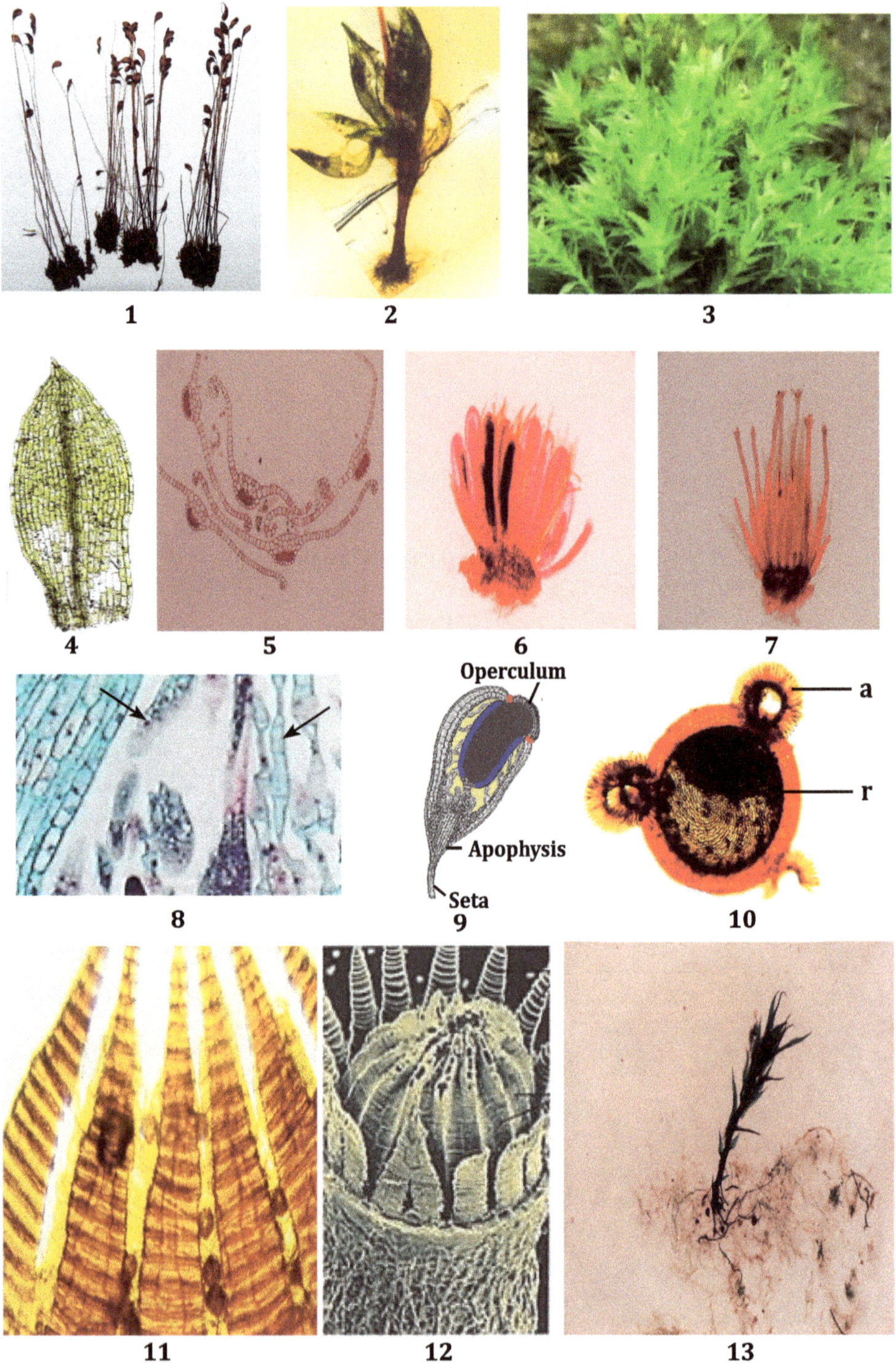

1 2 3

4 5 6 7

Operculum
Apophysis
Seta
a
r

8 9 10

11 12 13

PLATE VII: *Anthoceros*

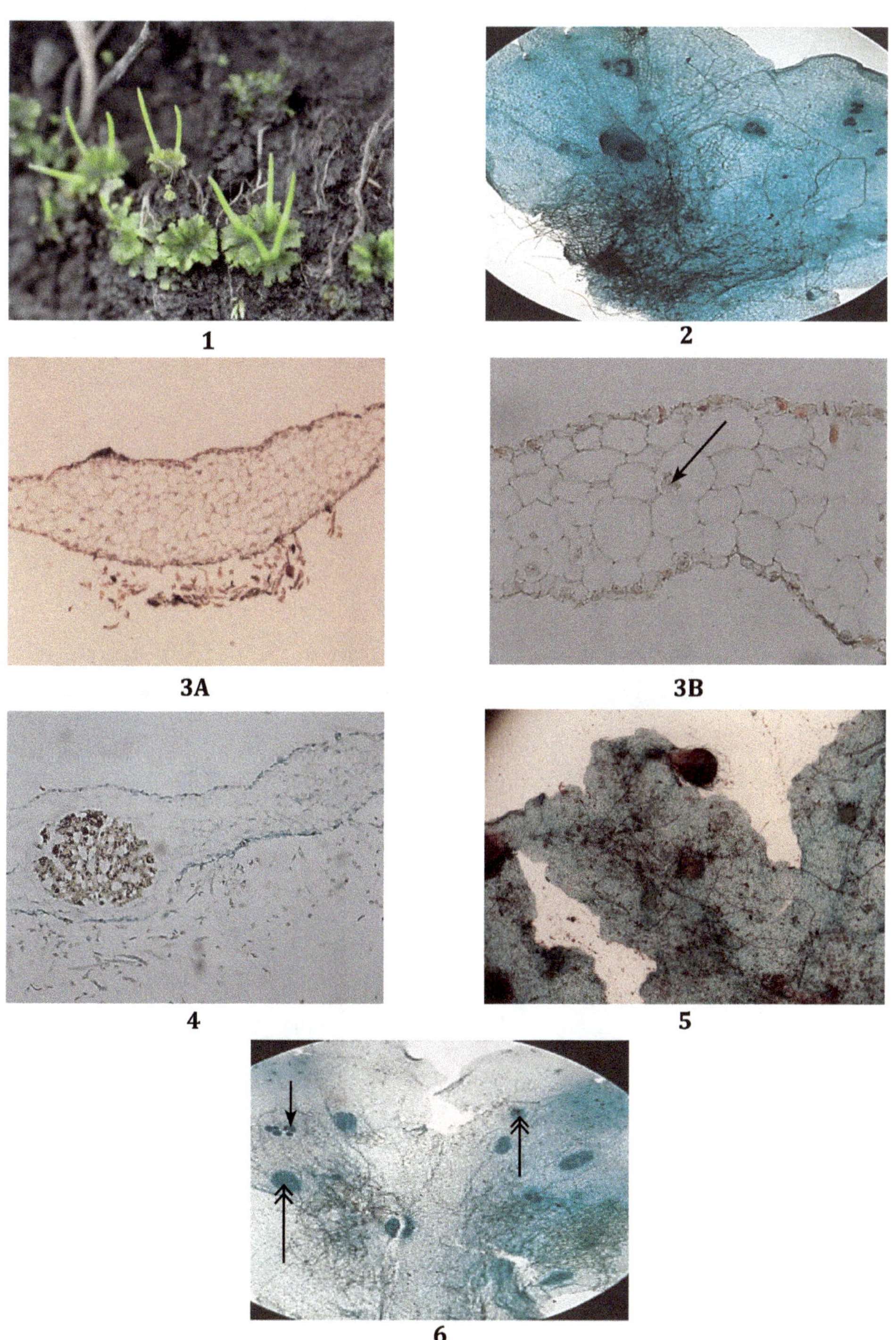

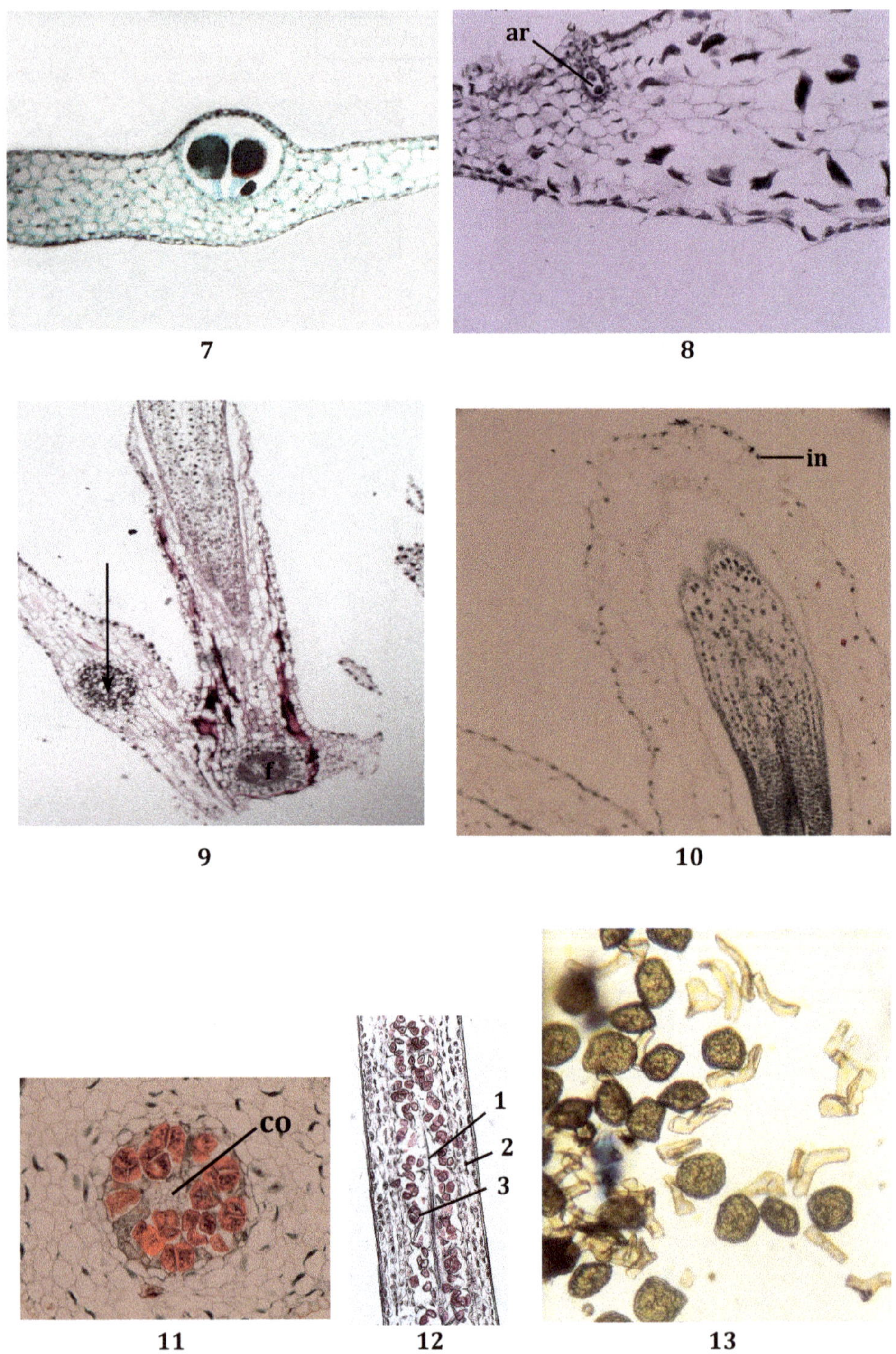

7 8 9 10 11 12 13

PLATE VIII: *Dispersal mechanisms in bryophyte*

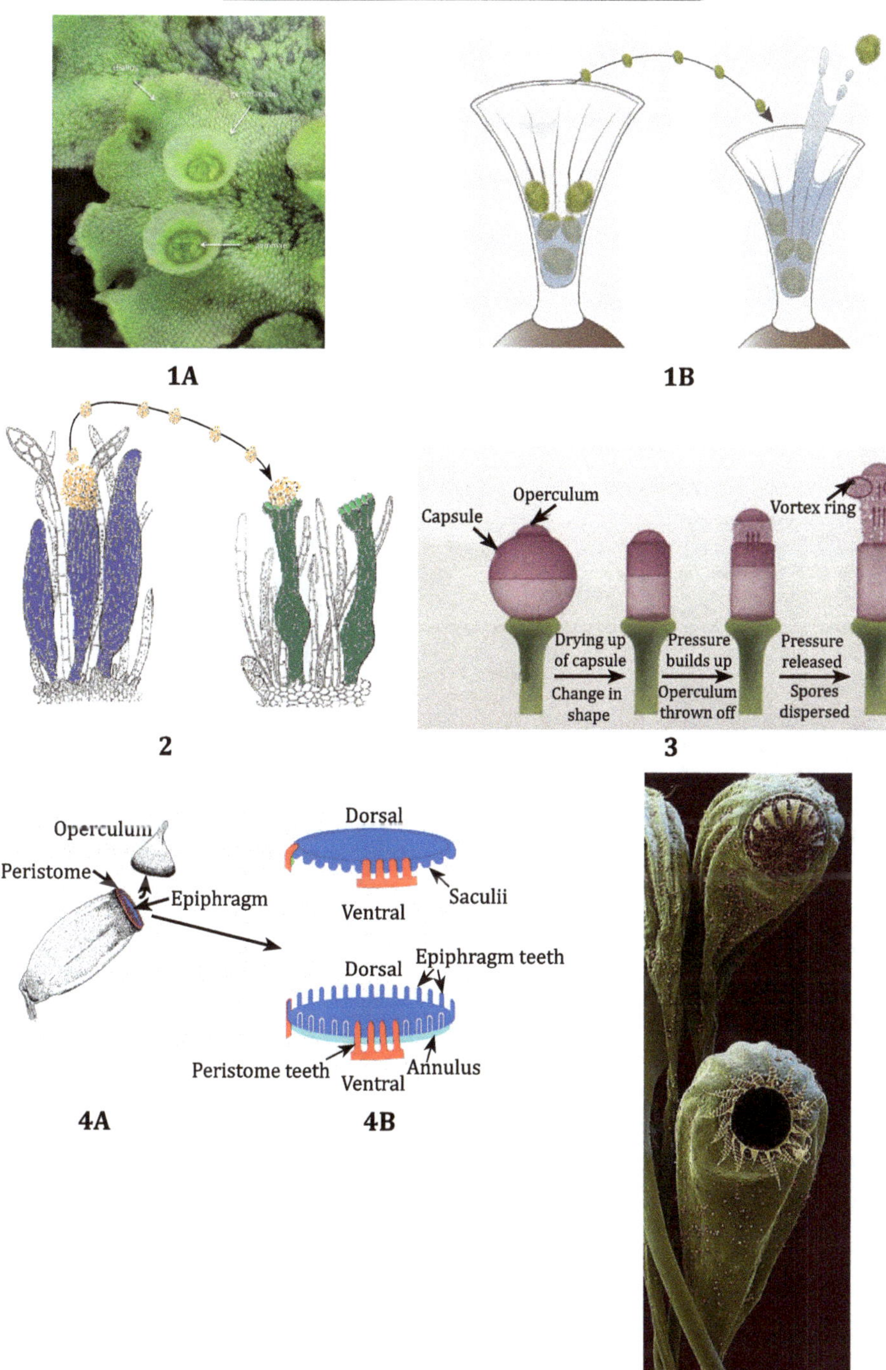

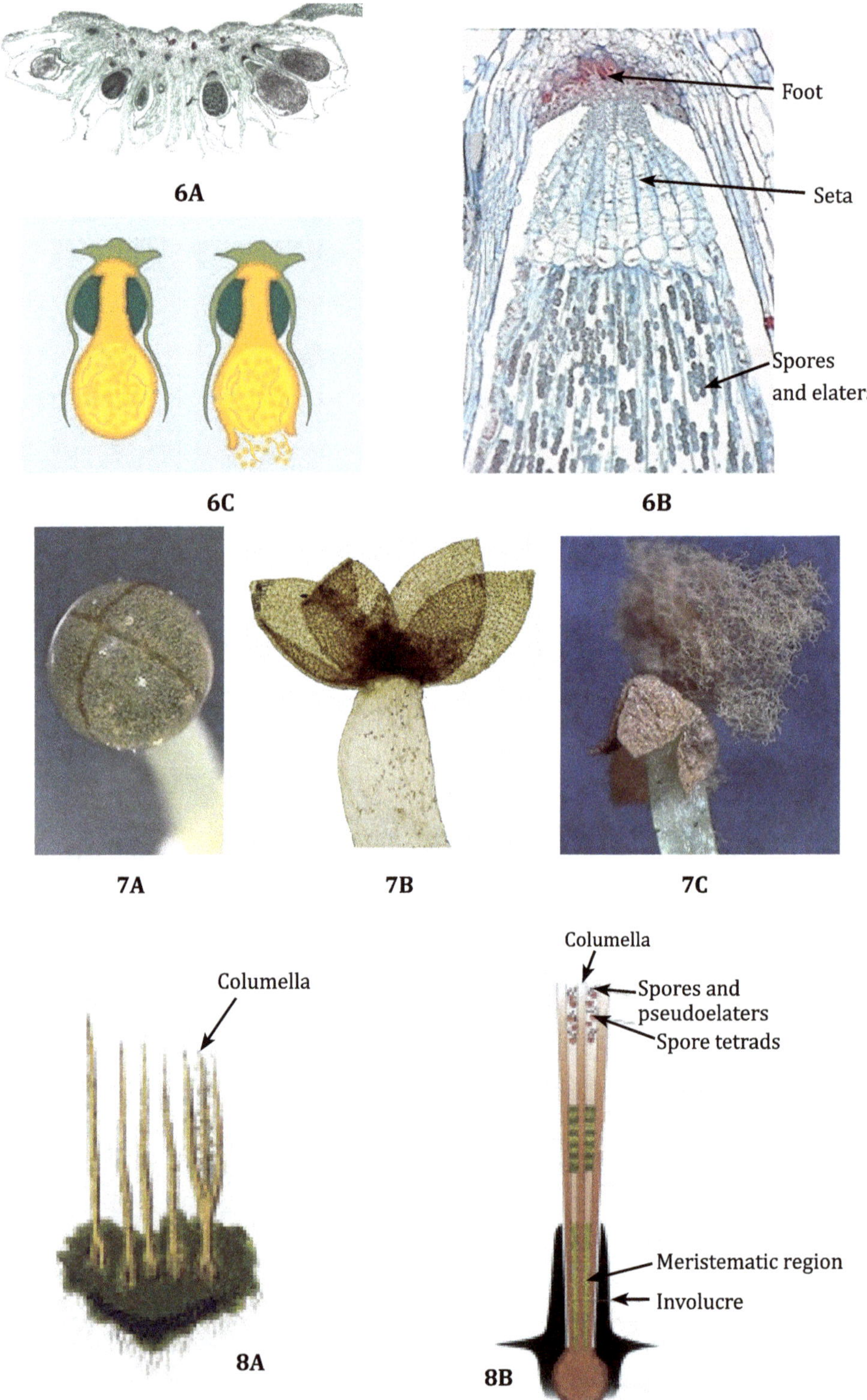

6A

6C

6B

7A

7B

7C

8A

8B

PLATE IX: *Some common liverworts for field study*

1

2

3

4

5

6

7

8

9

10

PLATE X: *Some common mosses for field study*

Appendix

Glossary

Acropetal: A term which explains the growth or development base upwards, *e.g.*, sex organs in *Riccia* with oldest towards the base and youngest towards the apex.

Amphigastria: In leafy liverwort like *Porella*, there are three rows of leaves – two rows of lateral leaves and one of ventral leaves. These undersurface leaves are smaller or rudimentary and are known as amphigastria.

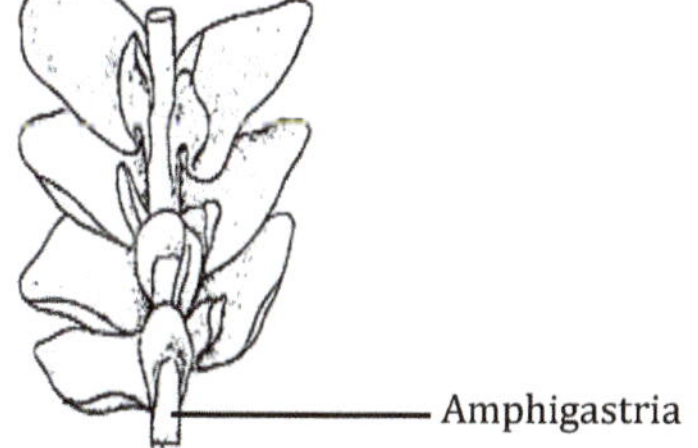

Amphithecium: Upon fertilization, a zygote is formed. The zygote develops into embryo which organizes two embryonic layers. The amphithecium (along with endothecium) which gets differentiated in early stages of development, generally gives rise to the wall of the capsule in liverworts while in mosses it may form wall and spore sac.

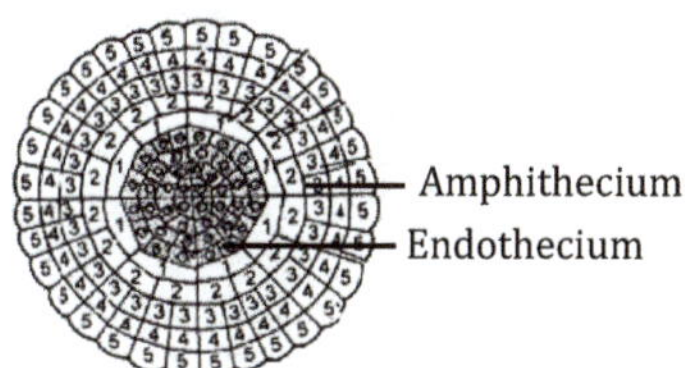

Antheridium: The male sex organ of algae, mosses, ferns, fungi, and other non-flowering plants such as bryophytes and pteridophytes, with a jacket that encloses antherozoids.

Archegoniophore/antheridiophore: It is the stalk or other outgrowth of a thallus upon which archegonia/antheridia are borne (as in liverwort *Marchantia*). Also referred to as carpocephalum or receptacle. It bears a stalk and a disc.

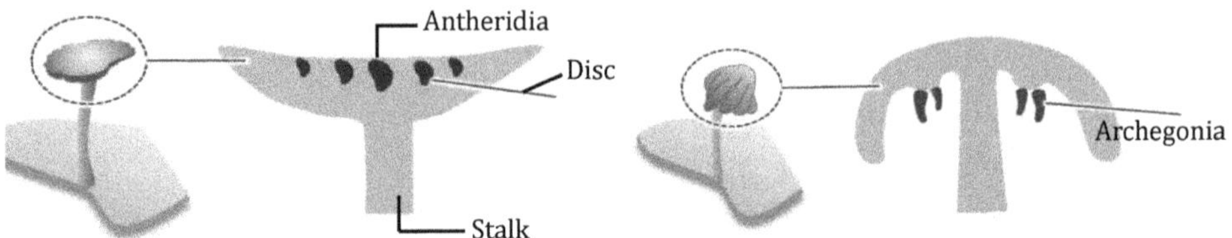

Archegonium: The flask-shaped female sex organ in archegoniates with a neck and a venter, and the egg present inside the venter. The neck has cover cells which are dislodged to facilitate antherozoid entry. An important structure which has brought bryo-, pterido- and gymnosperms together as a non taxonomic assemblage – the archegoniates.

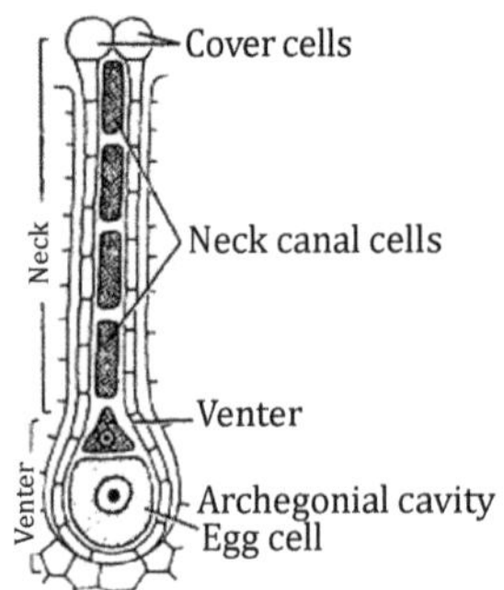

Archesporium: It is the fertile tissue (diploid) that develops inside the spore sac and undergoes development to form sporogenous mass which through meiosis forms spores (haploid) as in bryophytes. However, archesporium is also seen in pollen grain formation in gymnosperms and angiosperms.

Arthodontous mosses: Mosses with a peristome in which the teeth are structured by articulated cell wall remnants and where each tooth is composed of three concentric peristomial cell layers: the outer, primary, and inner, peristomial layers (OPL, PPL, and IPL). Often, a double ring of teeth is present - external ring of teeth contributes to the **exostome** while inner contributes to **endostome**.

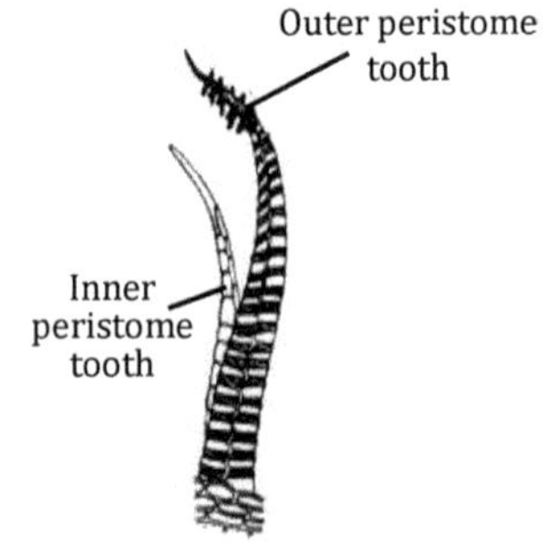

Bryophytes: Refers to a group of plants that includes, liverworts, mosses and hornworts. All of these are nonvascular plants (under embryophyta) with gametophyte as the dominant phase. They are also called amphibians of the plant kingdom because for sexual reproduction they depend on water. They are considered as the first land plants which gave rise to the seed bearing groups of plants through evolution.

Calyptra: It is a post-fertilization haploid structure arising from the venter of archegonium in bryophytes. It is a protective layer around embryonic sporophyte and is well defined in advanced mosses where the sporophytic growth is influenced by calyptra. The **calyptra** is usually lost before the spores are released from the capsule. In several mosses it may stay like a protective cap or hood, covering the capsule.

Capsule: Is a part of the sporophyte which contains spores. The capsule may be simple (*Riccia*) or complex (*Funaria*). In *Riccia* it is equated with sporogonium/ sporophyte and is a sac-like structure that encloses spores while in *Funaria*, it has apophysis, theca and an operculum. In hornworts the entire sporophyte encloses sporogenous tissue at different stages of development. There are various mechanisms for its dehiscence and spore dispersal.

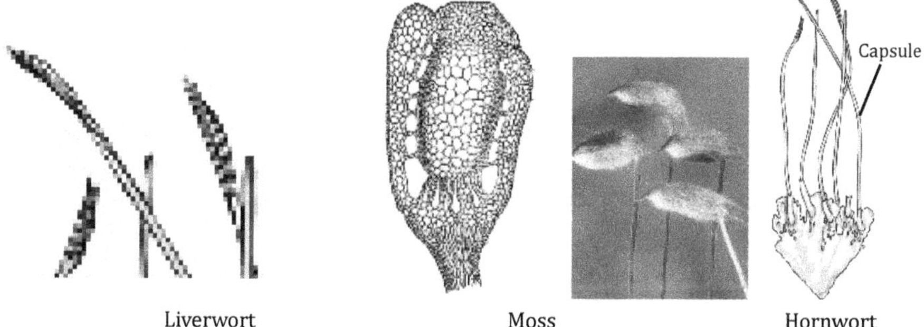

Carboniferous: A period of geologic time beginning nearly 345 million years ago, and lasting for nearly 65 million years.

Cretaceous: A period of geologic time beginning approximately 136 million years ago and lasting approximately 71 million years.

Dorsiventral: It is based on the architecture of the flattened structure as in thallus or leaf with distinct upper (dorsal) and lower (ventral) surfaces.

Elater: An **elater** is a sterile cell (or structure attached to a cell) in capsule, is hygroscopic, and undergoes changes like coiling and uncoiling or twisting and untwisting in response to changes in moisture in the environment. These movements help in spore dispersal. Elaters add to the sterility of the sporogenous tissue but aid in spore dispersal.

Elaterophore: A group of sterile cells bearing elaters, found in some liverworts such as *Pellia*. They may arise at the base of the capsule or at the apex.

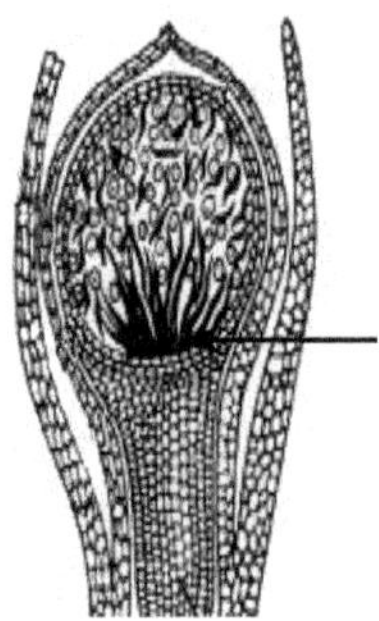

Gametophyte: The gamete- producing and usually a haploid phase in the life cycle of plants with alternating generations, and which leads to zygote formation upon fertilization. In bryophyte life cycle, this is the dominant phase.

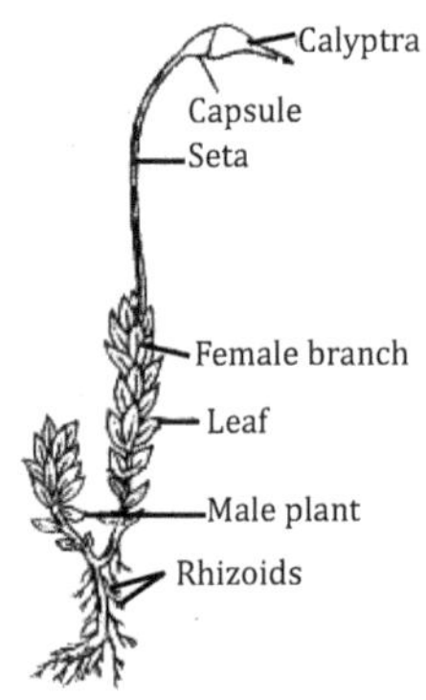

Gametophore: The stage in the gametophyte that bears the sex organs. The spore germinates to form a protonema which may be distinct or indistinct; it may be ephemeral or long lived. This is the first structure in the gametophytic phase of the plant and it gives rise to leafy long lived gametophore in mosses.

Gemma: Is a vegetative structure produced on any part of the thallus (rhizoids, stem), is disseminated and used as a propagule which gets detached from the parent and develops into a new individual. It is mode of vegetative reproduction.

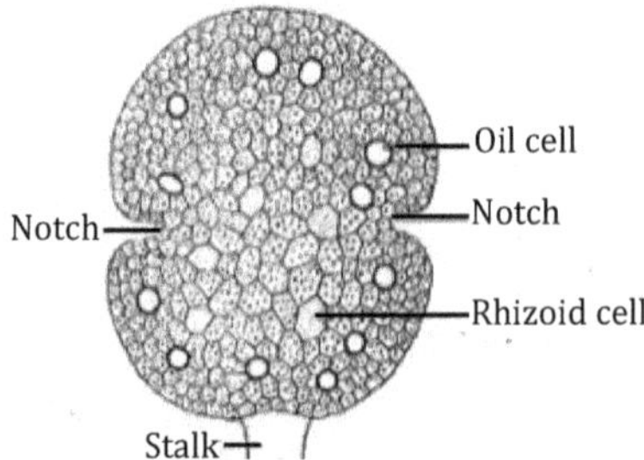

Hydrome: Tissue formed of hydroids, a type of vascular cells that occur in certain advanced bryophytes. Collectively, hydroids function as a conducting tissue, known as the **hydrome**, transporting water and minerals drawn from the soil.

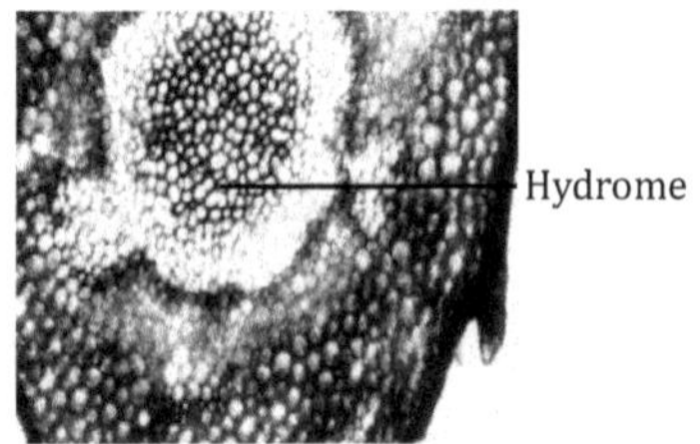

Incubous: It is the arrangement of leaves in a liverwort such as *Porella* where leaves are arranged with their upper edges overlapping the lower edges of the leaves above. The opposite of ***incubous*** is succubous.

Involucre: It may be a membranous envelope arising from the thalloid gametophyte as in *Pellia* and *Anthoceros* or a structure made from leaves as in *Porella*. It gives protection to the sporophyte.

Leptome: Plant tissue that conducts food similar to phloem in higher plants. It is formed from cells known as leptoids.

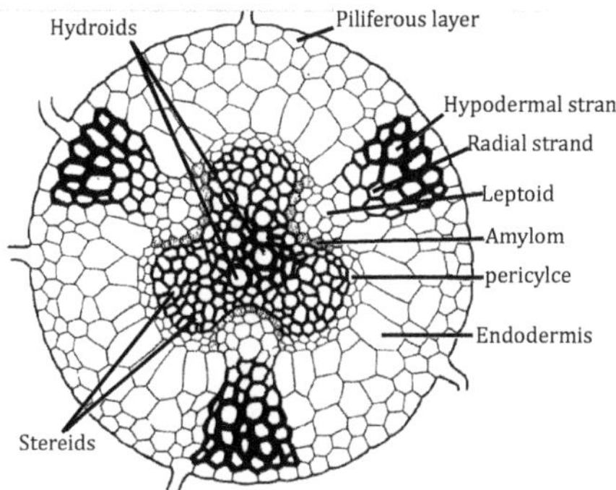

Nematodontous mosses: Nematodontous peristome teeth consist of whole dead cells and are evenly thickened as in *Polytrichum*. They do not show any ornamentations.

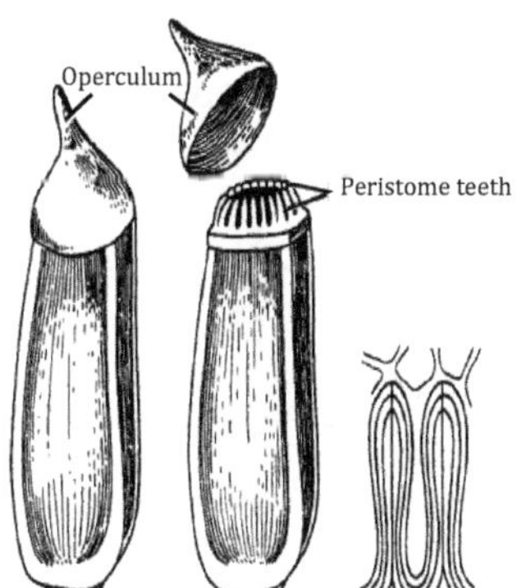

Operculum: Is a lid-like structure present in the capsule of mosses (and in other bryophytes where the antheridial opening may said to have special cells for dehiscence). When the operculum is thrown off, the spores from capsule (and antherozoids from antheridium) are released.

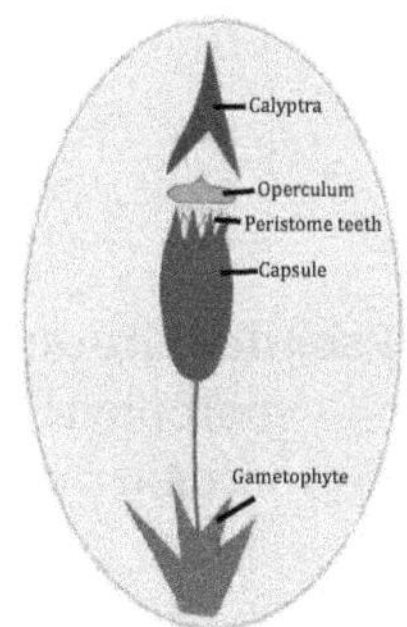

Peat: Accumulated mass of partially decomposed plant material, mainly of *Sphagnum*. It is of great commercial significance.

Perichaetium: A cluster of modified leaves surrounding the sex organs generally the archegonium or later the seta of mosses associated with protection from desiccation as they form cup like structure. Also seen in *Marchantia* where it arises from the gametophyte disc covering the entire row of archegonia.

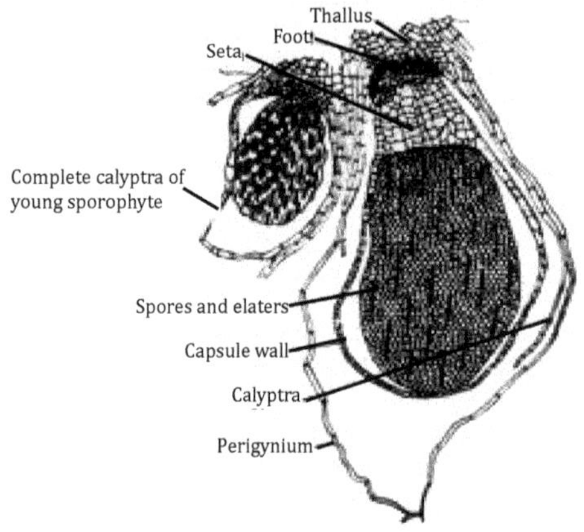

Permian: A period of geologic time beginning approximately 280 million years ago and lasting 55 million years.

Plagiotropic: Growth of a plant part at an oblique angle to a stimulus, such as gravity as seen in *Marchantia.*

Pleurocarpous: A moss in which the archegonia are borne on lateral shoots by subapical cells; thus the sporophytes are laterally placed.

Protonema: Is a filamentous green structure formed upon spore germination of a bryophyte. It brings bryophytes close to algae. Protonema may be ephemeral or long-lived and has a caulonemal and chloronemal system in mosses.

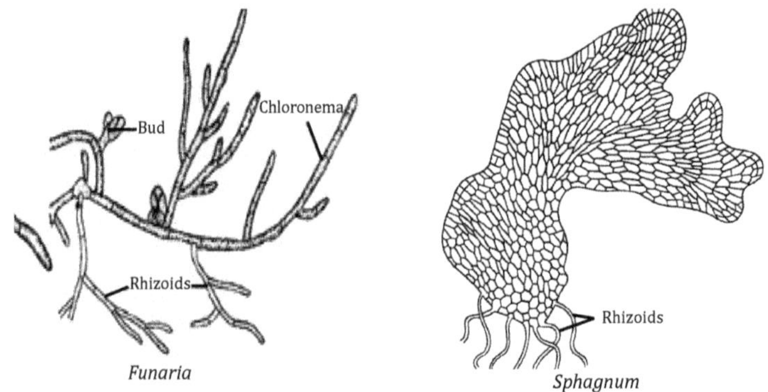

Funaria
Sphagnum

Pseudoelaters: Moisture-sensitive small and branched cells produced in the sporophyte of hornworts. They are chlorophyllous when young but as spores become ready for release, they help in spore dispersal.

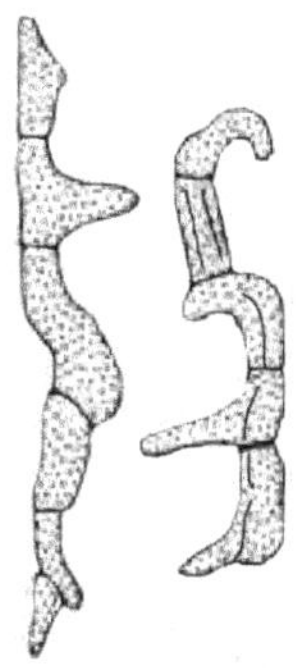

Rhizoids: Simple, hair-like projections that arise from the lower epidermal cells of the thallus in lower plants as in bryophytes, serving both to anchor the plant and (in terrestrial forms) to conduct water as in higher mosses such as *Polytrichum* where they form coiled structure for extra-capillary conduction.

Scales: Are membranous structures present on ventral surface of liverworts, generally present along midrib are rich in anthocyanin. They are protective in function.

Seta: Refers to the stalk supporting the capsule of a moss or liverwort and supplying it with nutrients. The seta connects the foot of the moss with the capsule. Besides conduction the seta may assist in spore dispersal. For illustration see sporogonium.

Sporogonium: In mosses and liverworts, the sporophyte generation that develops after sexual reproduction and produces spores. The term sporogonium is generally equated with sporophyte, but while sporophyte is a phase in life cycle marked with alternation of generation, sporogonium is a structure which in advanced members is made up of an absorptive *foot*, a stalk (*seta*), and a spore-producing *capsule*.

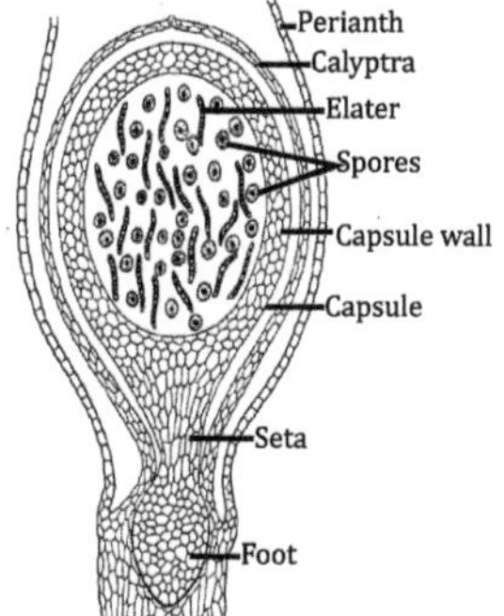

Sporophyte: Is the asexual and diploid phase in the life cycle of an organism, known to produce spores through meiotic divisions from which the gametophytic phase arises. In bryophytes, the phase starts with zygote formation and continues till spores are formed. It is the dominant form in vascular plants such as pteridophytes, gymnosperms and angiosperms.

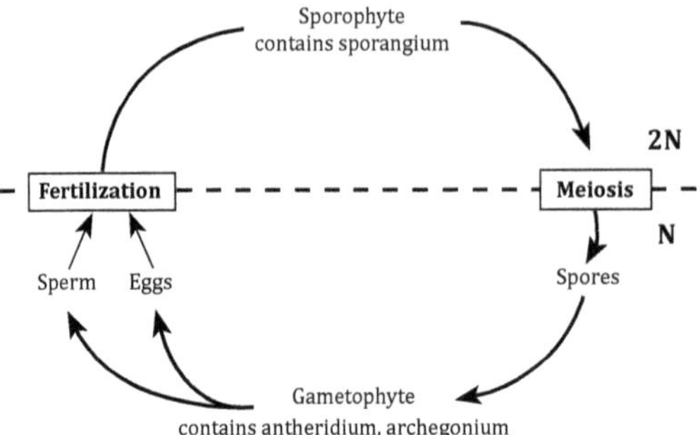

Sterome: The major part of central cylinder in mosses is made up of elongated cells which form the sterome, fortified with sclerenchyma.

Triassic: A period of geologic time beginning 225 million years ago and lasting 35 million years.

Turf: Thickly matted; used to describe gametophores that are densely crowded together with upright shoots.

Venter: The expanded portion of the archegonium that encloses the egg. Upon fertilization, it gives rise to calyptra, an important tissue for complete growth of the sporophyte. For illustration see archegonium.

Suggested Readings

Alaba, S. *et al.* 2015. The liverwort *Pellia endiviifolia* shares microtranscriptomic traits that are common to green algae and land plants. New Phytol. 206(1): 352–367.

Anderson, L.E. 1954. Hoyer's solution as a rapid permanent mounting medium for bryophytes. Bryologist 57: 242-244.

Asakawa, Y. 2004. Chemosystematics of the Hepaticae. Phytochemistry 65: 623–669.

Bahuguna, Y.M., Gairola, S., Semwal, D.P., Uniyal, P.L. and Bhatt, A.B. 2012. Bryophytes and Ecosystem. In: Paliwal, G.S., Kumar, M., Gupta, R.K. (eds.). Biodiversity of Lower Plants, I.K. International Publishing House Pvt. Ltd, New Delhi, India.

Bapna, K.R. and Kachroo, P. 2000. Hepaticology in India. Himanshu Publications, Udaipur, New Delhi. 2: 381-458.

Bell, P.R. and Hemsley, A.R. 2000. Green Plants: Their Origin and Diversity. Cambridge University Press, UK.

Buch, H. 1947. Ueber die Wasser und Mineralstoffversorgung der Moose. II. Commentat. Biol. 20:1-61.

Burgeff, H. 1943. Genetische Studien an *Marchantia*. Gustav Fischer, Jena. Cambridge University Press.

Chopra, R.S. 1975. Taxonomy of Indian Mosses. CSIR, New Delhi, India.

Clymo, R.S. and Hayward, P.M. 1982. The Ecology of *Sphagnum.* In: Smith, A.J.E (ed). Bryophytecology. Chapman & Hall, London.

Corliss, J.O. 1998. Haeckel's Kingdom Protista and Current Concepts in Systematic Protistology. Stapfia 56. Zugleich Kataloge des OÖ. Landesmuseums, Neue Folge Nr. 131: 85-104.

Crandall-Stotler, B., Stotler, R.E. and Long, D.G. 2009. Phylogeny and classification of the Marchantiophyta. Edinburgh J. Bot. 66:155-198.

Dandotiya, D., Govindpyari, H., Suman, S. and Uniyal, P.L. 2011. Checklist of the Bryophytes of India. Arch. Bryol. 65:1-117.

Daniels, A.E.D. and Kariyappa, K.C. 2007. Bryophyte diversity along a gradient of human disturbance in the southern Western Ghats. Curr. Sci. 93(7): 976- 982.

Durand, E.J. 1908. The development of the sexual organs and sporogonium of *Marchantia polymorpha*. Bull.Torrey Bot. Club. 35(7): 321-335.

Fremstad, E., Paal, J. and Möls, T. 2005. Impact of increased nitrogen supply on Norwegian lichen-rich alpine communities: a ten-year experiment. J. Ecol. 93:471- 481.

Frey, W. and Stech, M. 2005. A morpho-molecular classification of the Anthocerotophyta (Hornworts). Nova Hedw. 80: 541- 546.

Gangulee, H.C. 1976. Mosses of Eastern India and Adjacent Regions. 2: 1135-1140. Books and Allied Limited, Calcutta, India.

Gilley, S. 1982. The non-game update: the Delmarva fox squirrel; making a comeback? Virginia Wildlife 43(12): 24-25.

Goffinet, B. and Shaw, A.J. 2009. Bryophyte Biology. 2nd edition. Cambridge University Press, Cambridge.

Govindapyari, H., Kumari, P., Bahuguna, Y.M. and Uniyal, P.L. 2012 Evaluation of species richness of acrocarpous mosses in Imphal District, Manipur, India. Taiwania. 57(1): 14-26.

Groth-Malonek, M., Pruchner, D., Grewe, F., and Knoop, V. 2005. Ancestors of trans-splicing mitochondrial introns support serial sister group relationships of hornworts and mosses with vascular plants. Molec. Biol. Evol. 22: 117- 125.

Grytnes, J.A., Heegaard, E. and Ihlen, P.G. 2006. Species richness of vascular plants, bryophytes, and lichens along an altitudinal gradient in western Norway. Acta Oecologica. 29: 241- 246.

Hanson, D.T., Swanson, S., Graham, L.E., and Sharkey, T.D. 1999. Evolutionary significance of isoprene emission from mosses. Amer. J. Bot. 86: 634-639.

Hébant, C.1977. The Conducting Tissues of Bryophytes. J. Cramer, Lehre, Germany, 157 pp. + 80 Plates.

Hooper, D.U., Adair, E.C., Cardinale, B.J., Byrnes, J.E.K., Hungate, B.A., Matulich, K.L., Gonzalez, A., Duffy, J.E., Gamfeldt, L., and O'Connor, M.I. 2012. A global synthesis reveals biodiversity loss as a major driver of ecosystem change. Nature 486:105-108.

Jian-Guo, C., Xi-Ling, D., Hong-Mei, Z., and Quan-Xi, W. 2014. Formation and development of rhizoids of the liverwort *Marchantia polymorpha*. J.Torrey Bot. Soc. 141:126-134.

Lang, W.H. 1901. On apospory in *Anthoceros laevis*. Ann. Bot.15: 503-510.

Lorbeer, G. 1934. Die zytologie der lebermoose mit besonderer berucksichtingung allgemeiner chromosomenfragen. Jahrb.Wiss. Bot. 80:567-817.

Mc Conoha, M. 1941.Ventral structures effecting capillarity in the Marchantiales. Am. J. Bot. 28:301-306.

Meyer, H. and Santarius, K.A. 1998. Short-term thermal acclimation and heat tolerance of gametophytes of mosses. Oecologia 115: 1-8.

Muggoch, H. and Walton, J. 1942. On the dehiscence of the antheridium and the part played by surface tension in the dispersal of spermatocytes in Bryophyta. Proc. R. Soc. Lond. [Biol].448-461.

Negi, H.R. and Gadgil, M. 1997. Species diversity and community ecology of mosses: A case study from Garhwal Himalaya. J. Ecol. Environ. Sci. 23: 445-462.

Negi, H.R. and Gadgil, M. 2001. Ecological niche of certain terricolous liverworts from selected localities of Garhwal Himalayas: A preliminary study. In: Nath, V. and Asthana, A.K. (eds). Perspectives in Indian Bryology. Bishen Singh Mahendra Pal Singh, Dehradun, pp 23-33.

Negi, H.R. and Gadgil, M. 2002. Cross-taxon surrogacy of biodiversity in the Indian Garhwal Himalaya. Biol. Conserv. 105: 143–155.

Okada, Y. *et al.* 2000. Assignment of functional amino acids around the active site of human DNA topoisomerase II alpha. J. Biol. Chem. 275(32):24630-8.

Parihar, N.S.1965. An Introduction to Embryophyta. Volume 1, Bryophyta. Publisher: Central Book Depot; 5th edition.

Perold, S.M. 1995. Studies in the Marchantiales (Hepaticae) from southern Africa. 9, The genus *Marchantia* and its five local species. Bothalia 25: 2:727.

PickettHeaps, J. 1975. Green Algae. Structure, Reproduction and Evolution in Selected Genera. 606 S., 44 Strichzeichnungen, 882 Mikroaufnahmen. Sunderland, Mass. Sinauer Associates, Inc. Pbl./W. H. Freeman & Co.

Proctor, M.C.F., Ligrone, R. and Duckett, J.G. 2007. Desiccation tolerance in the moss *Polytrichum formosum*: Physiological and fine-structural changes during desiccation and recovery. Ann. Bot. 99:75–93.

Proctor, M.C.F., Oliver, M.J., Wood, A.J., Alpert, P., Stark, L.R., Cleavitt, N.L., and Mishler, B.D. 2007. Desiccation-tolerance in bryophytes: A review. Bryologist 110:595–621.

Qiu, Y.L., Li L, Wang B, et al. 2007. A nonflowering land plant phylogeny inferred from nucleotide sequences of seven chloroplast, mitochondrial, and nuclear genes. Int. J. Plant Sci. 168:691–708.

Qiu, Y.L., Li, L., Wang, B., Chen, Z., Knoop, V., Groth-Malonek, M., Dombrovska, O., Lee, J., Kent, L., Rest, J., et al. 2006. The deepest divergences in land plants inferred from phylogenomic evidence. Proc. Natl. Acad. Sci. USA 103: 15511–15516.

Rabatin, S.C. 1980. The occurrence of the vesicular-arbuscular mycorrhizal fungus *Glomus tenuis* with moss. Mycologia 72:191-195.

Renzaglia, K.S., Villarreal, J.C. and Duff, R.J. 2009. New insights into morphology, anatomy and systematics of hornworts. In: Goffinet, B. & Shaw, A.J. (eds.), Bryophyte Biology, ed. 2. Cambridge, Cambridge University Press, pp. 139–171.

Schofield, W.B. 1985. Introduction to Bryology. MacMillan Publishing Company. NY.

Schwarzenbach, M.1926. Regeneration und Aposporie bei *Anthoceros*. Dissertation/Zürich. Sonderdruck aus: Archiv der Juliu Klaus-Stiftung für Vererbungsforschung, Sozialanthropologie und Rassenhygiene Zürich, Bd.II, Heft 2, S.91-141., Zürich, Orell Füssli.

Sironval, C.1947. Experience sur les stages de developement de la forme filamentewe en culture de *Funaria hygrometrica*. L. BuU. Soc. Bot. Belg.79: 48-78.

Smith, G.M.1955. Cryptogamic Botany. Vol. II. Bryophytes and Pteridophytes. New York: McGraw-Hill.

Stotler, R.E. and Crandall-Stotler, B. 2005. A revised classification of the Anthocerotophyta and a checklist of the hornworts of North America, north of Mexico. Bryologist 108: 16-26.

Thomas, R.J., Stanton, D.S., Longendorfer, D.H. and Farr, M.E. 1978. Physiological evaluation of the nutritional autonomy of a hornwort sporophyte. Bot. Gaz. 139: 306-311.

Tanwir, M., Langer, A. and Bhandari, M. 2008. Liverwort and hornwort flora of Patnitop and its adjoining areas (J&K), Western Himalaya, India. Geophytology 37.

Van Tooren, B.F., 1988. The fate of seeds after dispersal in chalk grassland: the role of the bryophyte layer. Oikos 53: 41-48.

Villarreal, J.C., Cargill, D.C., Hagborg, A., Söderström, L. and Renzaglia, K.S. 2010. A synthesis of hornwort diversity: patterns, causes and future work. Phytotaxa 9: 150-166.

Villarreal, J.C., and Renner, S.S. 2012. Hornwort pyrenoids, carbon-concentrating structures, evolved and were lost at least five times during the last 100 million years. Proc. Natl. Acad. Sci. U.S.A. 109: 18873-18878.

von Konrat, M. and Braggins, J.E. 2001a. A taxonomic assessment of the initial branching appendages in the liverwort genus *Frullania* Raddi. Nova Hedwigia 72: 283-310.

von Konrat, M. and Braggins, J.E. 2001b. Notes on five *Frullania* species from Australia, including typification, synonyms, and new localities. J. Hattori Bot. Lab. 91: 229-263.

Zechmeister, H.G., Schmitzberger, I., Steurer, B., Peterseil, J. and Wrbka, T. 2003. The influence of land-use practices and economics on plant species richness in meadows. Biol. Conserv. 114 (2): 165-177.

Internet Sources

https://en.wikipedia.org/wiki/Süsswassertang

http://starwars.wikia.com/wiki/Planetary_Trade_Director

https://books.google.co.in/books?id=CDwpCgAAQBAJ&pg=PT68&lpg=PT68&dq)

https://en.wikipedia.org/wiki/Vivo_per_lei

https://patents.google.com/patent/US5718697A/en

https://www.anbg.gov.au/bryophyte/sex-sperm-dispersal.html).

Question Bank

Q1. Name a bryophyte with: ***Admissions 2004 onwards***

i) Vaginula

ii) Pyrenoids

iii) Elaterophore

iv) 16 peristomial teeth in a ring

v) Isobilateral spores

vi) Elaterophore

vii) Pseudoelaters

viii) Aquatic habitat

ix) Retort cells

x) Leptome

xi) Embedded sporophyte

xii) Rosette habit

xiii) Bog moss

xiv) Meristematic region in sporophyte

Q2. Draw a well labelled diagram of:

i) L.S. mature archegoniophore-*Marchantia.*

ii) L.S. capsule-*Polytrichum*

iii) V.S. leaf-*Polytrichum*

iv) T.S. capsule-*Anthoceros* at the level of spores.

v) L.S. sporophyte-*Funaria*

vi) L.S. archegonial head-*Funaria.*

vii) L.S. sporophyte-*Anthoceros*

viii) L.S. sporophyte-*Pellia* (*2013*)

ix) L.S. capsule-*Funaria* (2014)

Q3. *Answer in detail*

i) Progressive sterilization of sporogenous tissue in bryophytes is accompanied by efficient method of spore dispersal. 1990

ii) Discuss the ecological and economic significance of *Sphagnum*. 1990

iii) Justify that the sporophyte of *Riccia* is a new individual and not an outgrowth of the thallus.

OR

Is it appropriate to designate the mature sporophyte of *Riccia* a sporophyte in the strict morphological sense? Explain.

Semester mode

iv) Dehiscence and spore dispersal in *Sphagnum.*

v) Compare the organization of the operculum region in the capsule of *Funaria* and *Polytrichum*. Which according to you has a more efficient spore dispersal.

vi) Compare the thalli of *Pellia* and *Porella.* (2013)

v) Compare spore dispersal mechanism of *Riccia, Marchantia, Anthoceros* and *Pellia.* (2013)

vi) Why are bryophytes regarded as amphibians of Plant Kingdom?

vii) Discuss adaptations of bryophytes for land habit. (2014)

vii) Compare sporophytes of *Marchantia* and *Anthoceros.* (2014)

Q4. *Write short notes on:*

i) Bryophytes as plant indicators of pollution. (1990)

ii) Structure of gemma cup of *Marchantia.*

Semester mode

iii) Asexual mode of reproduction in *Marchantia.*

Semester mode

iv) Anatomical features of gametophyte of *Sphagnum.*

Semester mode

v) Moss protonema. Semester mode

vi) Vegetative propagation of bryophytes. (2013)

vii) Ecological significance of bryophytes. (2013)

Q5. Comment on:

i) *Anthoceros* is a synthetic type.

ii) Bryophytes are amphibians of plant kingdom.

iii) Mosses are better suited to land habit as compared to liverworts.

Semester mode

iv) Sporophyte of *Funaria* is partially an independent phase.

Semester mode

v) *Sphagnum* is a synthetic genus.

Q6. Distinguish between: **1990**

i) Elaters and pseudoelaters

ii) Apophysis and pseudopodium

iii) Perigynium and perichaetium

iv) Peristome of *Funaria* and *Polytrichum*

v) Gametophyte of *Porella* and *Anthoceros*

vi) Male and female receptacle of *Marchantia* (2013)

vii) Rhizoids of liverworts and mosses (2014)

viii) Leaves of *Porella* nd *Sphagnum* (2014)

ix) Thallus of *Marchantia* and *Pellia*. (2014)

Q7. Define: **Admissions 2004 Onwards**

i) Peristome

ii) Elaterophore

iii) Annulus

iv) Appendiculate scales

v) Protonema

vi) Tubers

vii) Gemma

viii) Incubous arrangement of leaves.

ix) Succubous arrangement of leaves.

x) Retort cells

xi) Apophysis

Index

R

S

T

V

www.ingramcontent.com/pod-product-compliance
Ingram Content Group UK Ltd.
Pitfield, Milton Keynes, MK11 3LW, UK
UKHW021011290726
14059UKWH00001BA/71